POCKET PROGRAMMABLE CALCULATORS IN BIOCHEMISTRY

Pocket Programmable Calculators in Biochemistry

JOHN E. BARNES
E. I. duPont de Nemours, Inc.
Wilmington, Delaware

ALAN J. WARING
Johnson Research Foundation
Philadelphia, Pennsylvania

A WILEY-INTERSCIENCE PUBLICATION

JOHN WILEY & SONS

New York · Chichester · Brisbane · Toronto

Copyright © 1980 by John Wiley & Sons, Inc.

All rights reserved. Published simultaneously in Canada.

Reproduction or translation of any part of this work
beyond that permitted by Sections 107 or 108 of the
1976 United States Copyright Act without the permission
of the copyright owner is unlawful. Requests for
permission or further information should be addressed to
the Permissions Department, John Wiley & Sons, Inc.

Library of Congress Cataloging in Publication Data:

Barnes, John E
  Pocket programmable calculators in biochemistry.

  "A Wiley-Interscience publication."
  Includes index.
  1. Biological chemistry—Problems, exercises,
  etc.—Data processing. 2. Programmable calculators.
  I. Waring, Alan J., joint author. II. Title.

QP518.5.B37   574.1'92'0285   79-2547
ISBN 0-471-06434-3
ISBN 0-471-04713-9 pbk.

Printed in the United States of America

10 9 8 7 6 5 4 3 2 1

This book is dedicated with much love to our wives, Susan and Pat

PREFACE

The programs for the Hewlett-Packard and Texas Instruments calculators presented in this volume are not intended to represent an exhaustive compilation of the types of problems encountered in biochemistry and molecular biology. Neither are these programs meant to represent necessarily the best methods to solve a particular problem. We have tried to present programs for the solution of problems which previously required the laboratory worker to either go to the university computing center or accept a less accurate solution. In some instances, more rigorous solutions to problems using computer programs not limited by memory are available. Often, however, the calculator solution is as rigorous as the computer solution and is much more accessible.

We wish to thank our colleagues for providing suggestions and criticisms as well as encouragement throughout the development of the calculator programs in this volume. We are grateful also to both Hewlett-Packard and Texas Instruments for providing us with photographs of calculators and/or other materials. Our special thanks go to Karl Marhenke, who provided us with an improved version of his HP-67 program for computing the titration curve of weak acid or base, and to others too numerous to mention, whose contributions to various journals and magazines provided us with routines and algorithms for our programs. We are indebted also to the following persons and organizations for useful ideas and information: Richard J. Nelson, publisher of *PPC Journal*, formerly *65-Notes*, and Richard Siegel and Ken Newcomer of the Hewlett-Packard Company.

We would sincerely appreciate your comments and criticisms, especially about any errors you may find.

We also wish to express our gratitude to our wives and families, without whose encouragement we might not have undertaken this project.

John E. Barnes
Alan J. Waring

Wilmington, Delaware
Philadelphia, Pennsylvania
January 1980

THE CALCULATORS

A. HEWLETT-PACKARD HP-67/97

The Hewlett-Packard HP-67 shown in Figure A is a reverse
Polish notation (RPN) card-programmable calculator capable of
storing and executing programs of up to 224 fully merged steps.
The term "fully merged" refers to the fact that any function on
the keyboard requiring one, two, or even three keystrokes requires only one step or line of program memory. There are also
26 data registers (16 primary and 10 secondary).

There are 10 user-definable keys for programming any special
functions that may be required, such as defining portions of a
program as subroutines. These subroutines may be executed from
the keyboard or from within a program. There are an additional
10 numeric labels (*LBL 0* through *LBL 9*) which may also be executed from the keyboard or from within a program.

The *pause* function interrupts program execution and displays
the current result for about 1 sec. During that 1 sec pause,
data can be entered from the keyboard or magnetic cards can be
read. This "active" pause is a most powerful and useful feature
and is used to good advantage in several programs in this book.

Either from the keyboard or from within a program, one has
the power to transfer (or branch) program execution to any other
part of the program. When followed by a label, the *GTO* key
branches directly to the specified label and continues execution.
When a series of instructions is executed several times in a program, program memory can be conserved by executing that portion
of the program as a subroutine. A *GSB* instruction followed by a
label branches program execution to the specified label (just as
does the *GTO* instruction); however, program execution is then

Figure A.   Hewlett-Packard HP-67.

"returned" automatically to the step following the *GSB* instruction when the next *RTN* is encountered . A *GSB* instruction can
be used within a subroutine (i.e., nested) to a depth of three
levels.

The conditional branching capability of the HP-67 allows a
program to make a decision by comparing the values in the X- and
Y-registers or by comparing the X-register with zero. If the
condition is true, the next instruction in program memory is
executed. (Remember the "Do if true rule.") If the condition is
false, the next step is skipped.

Flags can also be used for tests in programs. The four
flags available in the HP-67 can be set, cleared, or tested.
When a flag is tested, the calculator executes the next instruction ("Do if true"). The program skips the next step if the flag

is clear. Flags 0 and 1 are command-cleared; once they have been
set, they remain set until cleared. Flags 2 and 3 are test-
cleared; they are cleared automatically after a test and remain
clear until they are set again. Flag 3 is also a data entry-
sensing flag, that is, as soon as data are entered from the keyboard or from a magnetic card, flag 3 is set. This is one of the
most valuable features of the HP-67 and is used in several programs in this book.

The I-register can be used in several different ways in addition to the standard one, that is, for simple storage. The *(i)*
key combined with certain other functions uses the number stored
in the I-register to control those functions. This is referred
to as *indirect control*. The keys *GTO(i)* and *GSB(i)* perform a
direct branch or subroutine to a label specified by the current
number in the I-register. For example, if the I-register contains the number 3, a *GSB(i)* instruction performs a subroutine
to *LBL 3*. This is referred to as *indirect addressing*. When the
number in the I-register is negative, *GTO(i)* and *GSB(i)* instructions perform a direct branch or subroutine *backward* the number
of steps specified. This is referred to as *relative addressing*.
One can also use the I-register to specify the address of a
storage register. For example, *STO(i)* results in the displayed
number being stored in the storage register specified by the
value in the I-register. Similarly, *RCL(i)* recalls the contents
of the storage register specified by the value in the I-register.
Storage register arithmetic may also be performed on the contents of the register specified by the I-register. The *ISZ* and
DSZ instructions cause the I-register to be incremented or decremented, respectively, and then compared to zero. If the contents
of the I-register are equal to zero, the program skips the next
step in program memory. Instructions *ISZ(i)* and *DSZ(i)* cause the
contents of the register specified by the value in the I-register
to be incremented and decremented, respectively, and then compared to zero. Again, if the contents of the indirectly addressed register are equal to zero, the program will skip the next
step.

The contents of program memory and the data storage registers can be recorded on small magnetic cards and later re-ent-
ered. The so-called smart card reader "knows" whether the magnetic card contains data or program steps and loads the program
or data in the proper location without any further user inter-
vention. When recording programs, the HP-67 automatically re-
cords angular mode setting, display setting, and the status of
all four flags. To record or load all 224 steps or all 26 data
registers, both sides of the magnetic card must be read into the
calculator. If the second side of a magnetic card is required,
the calculator "prompts" by displaying "Crd" whether he or

she is recording or loading programs or data. And the user can
load either side first.

Editing programs is quite easy. To help find mistakes in a
program, the *SST* key can be used to execute one step at a time in
RUN mode. In *PRGM* mode, the *SST* key allows the user to single-
step through each instruction to compare the keycodes with the
program listing. The *BST* key is used to backstep one step at a
time. Insertion of operations can be accomplished by first pos-
itioning the program pointer at the step before the intended
insertion. Then the function or functions can be keyed in,
resulting in all subsequent instructions being "bumped" down one
step in program memory for each inserted operation. When *DEL*
is pressed, the displayed instruction is erased and all subsequent instructions are moved upward one step.

The HP-97 calculator is the printing version of the HP-67.
Although the keycodes for the various functions are different,
programs written and recorded on magnetic cards for one machine
can be loaded and executed on the other.

The preceding discussion is not meant as a substitute for
studying the owner's manual and programming guide. We urge the
calculator user to study the owner's manual thoroughly and work
the sample problems. The owner of an HP-67/97 may then be able
to use the programs in this book to aid in the development of
new programs to supplement these, as well as to alter or tailor
these programs to his or her particular requirements.

B. TEXAS INSTRUMENTS TI-58/59

The Texas Instruments TI-58/59 programmable calculators
shown in Figure B feature AOS^{TM}--an algebraic operating system
which allows the user to enter programs in a familiar, standard
mathematical format. The AOS^{TM} system features straightforward
left-to-right entry of an equation just as one would read or
write it in standard mathematical terms. Equations are evaluated
according to a built-in hierarchy which requires that multiply
and divide be executed prior to add and substract. All other
functions such as *log, sin, 1/x,* etc., immediately operate on the
value in the display. The order of execution of functions can be
altered using up to nine sets of nested parentheses allowing up
to eight pending operations.

The TI-58/59 gives the user the flexibility to vary the
allocation of storage capability between program steps and data
registers. At turn on, the TI-58 provides 240 program steps and
30 data registers. Repartitioning can provide a maximum of 480
program steps and zero data registers or zero program steps and
60 data registers. The TI-59 provides 480 program steps and 60
data registers at turn on and can be repartitioned to provide a

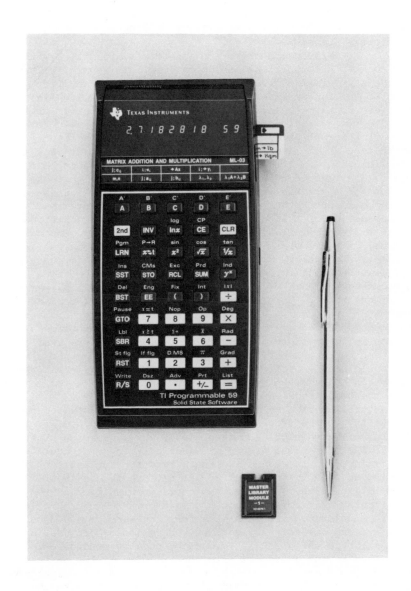

Figure B. Texas Instruments TI-59

maximum of 960 program steps, or up to 100 data registers and
160 program steps.

The TI-58/59 calculators are essentially identical with res--
pect to keyboard functions and operations. There are two types
of labels provided in the TI-58/59 calculators. The first consists of the ten user-definable keys (*A-E* and *A'-E'*) which can be
used for programming any special functions or defining portions
of a program as subroutines. These keys are often used for data
entry at those steps in a program that require data input. The
second type of label is the so-called "common" label. Any keyboard function (except *SST, DEL, LRN, Ind, 2nd,* and the numeric
keys 0 to 9) can be used as "common" labels to define those
portions of a program which would not usually be executed from
the keyboard but rather as subroutines within the program.

Branching or transfer instructions include unconditional
GTO instructions followed by a label or line number, and conditional branches where program transfer depends upon the outcome
of some sort of test. There are four types of display testing;
that is, the value in the display is compared to the value in the
independent test or "t"-register. These include $x \geq t$, $x \leq t$,
$x = t$, and $x \neq t$. The inverses of these relational tests can
also be programmed. Registers 00 through 09 can be automatically
incremented or decremented to control the number of times a subroutine is executed in a program loop. Ten flags are available
which can be set, reset, and tested to determine whether a transfer should occur. Following each kind of relational test there
must be an address, either a label or line number, which comp-
letes the conditional branching instruction.

Several addressing modes are featured which provide great
flexibility to the programmer. These include absolute (i.e.,
line number), indirect, and label modes for addressing program
steps. There are also direct and indirect modes for addressing
data registers. Instead of a single register for indirect addressing as used in the HP-67/97 all data registers in the
TI-58/59 can be used as index registers for indirect addressing.

Program editing is straightforward on the TI-58/59 calculator. Single step and back step functions are provided as well
as insert and delete. In addition, there is a "No-operation" key
(*NOP*) which is useful during program debugging procedures.

After keying a program into the calculator, the contents of
program memory and data registers can be recorded on blank magnetic cards with the exception of the TI-58 which lacks the
built-in card reader. An improved version of the TI-58 , desig-
nated the TI-58C has been recently introduced by TI. Featuring
continuous memory, the program is retained even when the calculator is turned off.

One of the most significant features of the TI-58/59 calculators consists of plug-in Solid State Software^{TM}. These library
modules contain up to 5000 prerecorded program steps which are
accessible to the user with only four keystrokes in the calculate mode or can be accessed as subroutines from within a user
written program. A master library module containing 25 programs
in the areas of math, statistics, finance, and general interest
subjects is included with each TI-58 and TI-59. Additional
Solid State Software^{TM} modules are available in areas of special
interest such as applied statistics, math utility routines, and
and *RPN* Simulator which may be of interest to the reader. Consult TI promotional literature for information on these and other
modules.

Another feature which provides significant capability is the
PC-100A thermal printer/plotter. Both the TI-58 and TI-59 can be
attached to this peripheral to obtain hard copy printouts of program listings and outputs. Editing is easier and faster using a
listing of the program for checking whether the instructions have
been keyed in correctly. In the *TRACE* mode, every calculation
and instruction is printed as it is executed. Subroutine calls
and returns can be followed as can all conditional and unconditional branches and loops. Using the special predefined operation (*Op*) codes the PC-100A can prompt by printing alphabetic
messages (20 characters per line), plot graphs, and print the
results of a calculation plus a four alpha character descriptor
of those results. The contents and the number of each register
and/or a list of all labels may also be printed.

The preceding discussion is not intended to substitute for a
thorough study of the owner's manual and programming guide. The
manual is quite detailed and should provide answers for any ques-
tions about the calculator and its operation. Additional sources
of information about TI calculators are listed in Appendix IV.

A WORD ABOUT
PROGRAM USAGE

Each program for Hewlett-Packard calculators in this volume
was listed using the printing HP-97 calculator with the exception
of the programs in Sections 2A and 6C. These programs were listed using the 82143A Peripheral Printer and HP-41C calculator
because an HP-97 was not available during the last stages of the
preparation of this book. The HP-41C owner will note that the
numeric labels preceding letter labels which are added during the
card reader translation process have been deleted. See Appendix
V. The first section of each subject area gives a description of
the problem and the equations used to solve it. We have not provided derivations of equations except when they are essential to
using or understanding the program. In most instances, derivations can be found in the references listed after the final
section ("Example"). In addition, we have attempted to provide
some analysis of the significance of the program, especially when
the program differs from older methods of data analysis. The
second section provides the instructions to enable the user to
enter the required data in the proper manner. The third section
contains listings of the program without the keycodes. In addition, the comments next to the steps describe the reasons for
the step, group of steps, or subroutine. These should provide
help in understanding the program flow and enable the user to
modify the program to meet his or her own needs. The fourth
section lists the data register contents and the labels for the
user-definable keys, which may be used to mark the blank magnetic
cards. The fifth section provides at least one sample problem
and the correct solution. Finally, references are provided to
to allow further reading. Print options have been included where
possible or appropriate for use with the HP-97 calculator.

   The listing format for the TI calculators is slightly different. The listings were made using the PC-100A printer at-
tachment for the TI-59. Print options are included in some of
the programs for the TI calculators. Using the alpha-numeric
print capability when connected to the PC-100A printer, both
input data and calculated results are labeled with a descriptor
of up to four characters long using Op code 6. Both the key
mnemonic and the keycode are listed for each program step. The
calculator status cannot be recorded on a magnetic card; therefore, the user must initialize his or her calculator according
to the required partitioning, and so on. The owner's programming manual should be consulted for instructions on properly
recording a program on a blank magnetic card.

   For HP calculators we have endeavored to utilize the symbols and conventions recommended and used by Hewlett-Packard.
These are described on page vi and vii of the Standard Pac. The
symbols for the TI calculators are different, and attempts have
been made to have them conform to the TI convention.

CONTENTS

POCKET PROGRAMMABLE
CALCULATORS
IN BIOCHEMISTRY

INTRODUCTION

For the professional biochemist in the laboratory, problem-solving and data analysis constitutes a day to day activity. One must not only solve trivial problems such as calculating the amount of a chemical to weigh out in order to make up a solution of specified molarity, but also perform repetitive calculations for large groups of data, reduce these data to a form which can be plotted, and determine the best fit parameters (e.g., slope and intercept for a straight line) which may then give a clearer understanding of the experimental results. Data reduction and analysis occupies a major portion of the scientist's time and takes valuable time away from the planning and carrying out of experiments. With the advent of the inexpensive pocket calculator, however, the time devoted to data analysis has been reduced considerably. Moreover, there has been a correspondingly significant increase in the reliability of the data themselves, since now relatively sophisticated statistical tests can be applied to determine the confidence limits and other measures of statistical significance. No longer must one labor for hours with a slide rule, tables, and pencil and paper. More time can now be devoted to planning and carrying out the experiment of interest.

The programmable pocket calculator is having an even more significant impact on the classroom and the laboratory than are the nonprogrammable models. Many problems can be solved with these devices that heretofore required the use of large computers in central university computing centers or of costly laboratory minicomputers. Moreover, the cost of such programmable calculators is decreasing and promises to decline even further as more units are sold and as the level of technology rises to allow

3

lower manufacturing costs. Hewlett-Packard and Texas Instruments are recognized as the current leaders in the design and manufacture of sophisticated programmable and nonprogrammable calculators. We have chosen the Hewlett-Packard HP-67/97 and the Texas Instruments TI-58/59 series of calculators as the machines upon which to implement the programs for solving complex problems of data reduction and analysis in biochemistry. These machines are the top-of-the-line models offered by these manufacturers. However, this does not mean that the programs and methods outlined in this book are limited to these machines.

Both Hewlett-Packard and Texas Instruments seem to be standardizing the calculator "languages" used in their calculators. Hewlett-Packard has implemented reverse Polish notation (RPN) with an operational stack in all of its scientific calculators. However, it has only been in the latest series of calculators that programs written for one HP calculator can, with little modification, be used for another. For example, programs written for calculators with less capacity, such as the HP-19C or HP-29C, can be used almost without modification on the HP-67/97 series. This reflects the upwardly compatibile nature of the calculator architecture, for example, fully merged keystrokes, conditional branches, and methods of register addressing.

Texas Instruments uses the algebraic operating system (AOS^{TM}), which is a combination of nested parentheses with pending operations and algebraic hierarchy; the order of execution is functions first (powers, roots, etc.) followed by multiply/divide and then by add/subtract. The newest series of TI calculators all use this system, including the TI-57 keystroke programmable calculator with 50 fully merged program steps. The upward compatibility of these machines illustrates once again that programs written for one machine are easily modifiable for use with another. Incidentally, the suggested retail price of the TI-57 is only $79.95, placing it well within the reach of the serious student of modern molecular biology and biochemistry. Although the HP-67/97 and the TI-58/59 calculators are somewhat more expensive, they too are becoming financially feasible for the serious student and the professional scientist in the laboratory.

Our intention is to provide brief descriptions of the types of problems encountered in the laboratory for day to day data acquisition and analysis, the equations and the methods used to solve them, commented keystroke listings, user instructions, and documentation. We hope that this approach will not only enable the professional scientist to get the most out of his or her calculator but also help the student to acquire a feel for the feasible in the design of experiments.

Although we have used programmable calculators to develop these programs, the nonprogrammable calculator is not disregarded. The equations and keystroke listings should enable the owners of nonprogrammable calculators to solve the same problems, albeit somewhat more slowly. In fact, many of the problems use the method of linear regression to obtain the best fit parameters of a straight line, and several nonprogrammable calculators which have linear regression as a "hard-wired" function are available for under $50. In short, our aim is to provide usable, well-documented calculator solutions to a wide variety of biochemistry problems for as wide an audience as possible.

I
AQUEOUS SOLUTIONS OF SMALL MOLECULES

1A. LABORATORY BUFFERS

Since most biochemical applications require relatively stable pH's, it is desirable to maintain the hydrogen ion concentration in solution by utilizing appropriate buffers. One property of a weak acid and its conjugate base (or, conversely, a weak base and its conjugate acid) is the capacity to resist pH change at the pK_a (i.e., where the ratio of weak acid to conjugate base is 1:1). Preparation of buffers for a specific pH range can be determined by using the Henderson-Hasselbalch expression, which is based on the equation for the dissociation constant of the weak acid HA:

$$K_a = \frac{[H^+] \ [A^-]}{[HA]} \tag{1}$$

By rearranging terms, taking logs, and multiplying both sides of equation 1 by -1, expression 2 is derived:

$$-\log[H^+] = -\log K_a - \log \frac{[HA]}{[A^-]} \tag{2}$$

Remembering that $pH = -\log[H^+]$ and pK_a, the familiar Henderson-Hasselbalch form of equation 2 is revealed:

$$pH = pK_a + \log \frac{[A]}{[HA]} \tag{3}$$

Relation 3 is valid assuming that the concentrations of weak acid and conjugate base are equal to their activities in solution (i.e., the ionization of HA is negligible in the presence of A^-).

The concentration-activity equivalence assumption holds as long
as the hydrogen ion concentration is less than one hundredth of
the weak acid concentration (this applies to most buffers in the
pH range of 3-9).

   Other important parameters which can affect buffer behavior
are the temperature and the ionic strength of the medium. Temperature changes affect the pK_a of the buffer and can be corrected by use of the temperature coefficient for the particular
buffer in question (Appendix III).

   The pK_a is also altered by the ionic strength of the medium.
Corrections in pK_a for ionic strength effects can be determined
by using equation 4 for weak acids and equation 5 for weak bases:

$$pK_a' = pK_a - (2n - 1)\left[\frac{0.5I^{\frac{1}{2}}}{1 + I^{\frac{1}{2}}} - 0.1I\right] \tag{4}$$

$$pK_a' = pK_a + (2n - 1)\left[\frac{0.5I^{\frac{1}{2}}}{1 + I^{\frac{1}{2}}} - 0.1I\right] \tag{5}$$

In equations 4 and 5, which are forms of the Debye-Hückel equation 3, I is the ionic strength (this is equivalent to concentration for weak acids or bases); n is the number of charges, and
pK_a is the pK_a under standard conditions.

   After correction of the pK_a for ionic strength effects, the
concentration ratio of the buffer components can be determined
using the Henderson-Hasselbalch equation:

$$\text{antilog } (pH - pK_a') = \frac{[A^-]}{[HA]} \tag{6}$$

$$\text{antilog } (pH - pK_a') = \frac{[B^+]}{[BA]} \tag{7}$$

User Instructions--RPN

Step	Instructions	Input	Keys	Output
1	Enter pK_a	pK_a	ENTER	
2	Enter number of charges	n	f a	
3	Enter buffer concentration	I	f b	
4	Enter temperature (°C); default is 25°C	T	f c	
5	Enter temperature coefficient of buffer	dpK_a/dT	f d	
6	Enter desired pH and calculate volume (ml) of HA^- or HA^{2-} required	pH	A	ml HA^-
7	Calculate volume (ml) of 1 M HA^{2-} required		R/S	ml HA^{2-}

Program Listing--RPN

Line	Key	Comments	Line	Key	Comments
001	*LBLa	Store n (number of charges)	019	*LBLd	Optional: Enter dpK_a/dT
002	STO1		020	STO4	
003	2		021	RCLB	
004	x		022	2	
005	1		023	5	
006	-		024	-	
007	STO3	→ $2n - 1$	025	x	
008	R↓		026	RCL0	
009	STO0	→ $pK_a^{25°C}$	027	+	Temperature correc-
010	RTN		028	STO0	ted pK_a
011	*LBLb		029	RTN	
012	STO8	→ I	030	*LBLA	
013	√x		031	STO6	Enter desired pH
014	STO2	→ $\sqrt{I}$	032	RCL2	
015	RTN		033	2	
016	*LBLc		034	÷	
017	STOB		035	RCL2	
018	RTN	Store T (°C)	036	1	

Line	Key	Comments
037	+	
038	.	
039	1	
040	RCL8	
041	x	
042	-	
043	÷	
044	STO4	
045	RCL3	
046	x	
047	RCL0	
048	X⇄Y	
049	-	
050	RCL6	
051	X⇄Y	
052	-	
053	10^x	
054	STO7	$10^{(\text{pH} - \text{p}K_a)}$
055	RCL1	
056	RCL3	
057	X≤Y?	
058	GTO1	
059	GTO2	
060	*LBL1	
061	RCL2	
062	RCL7	→ ml 1 M HA$^-$
063	÷	required to make
064	EEX	liter of buffer
065	3	
066	x	
067	RTN	
068	*LBL2	
069	RCL8	→ ml 1 M HA^{2-}
070	RCL7	required to make
071	3	liter of buffer
072	x	
073	1	
074	+	
075	÷	
076	STOA	
077	EEX	
078	3	
079	x	
080	R/S	
081	RCL7	
082	3	
083	x	

Line	Key	Comments
084	1	
085	+	→ ml 1 M HA
086	RCL7	required to make
087	RCL8	liter of buffer
088	x	
089	X⇄Y	
090	÷	
091	STO9	
092	EEX	
093	3	
094	x	
095	RTN	

Register Contents, Labels, and Data Cards--RPN

Register	Contents	Labels	Contents
R_0	pK_a	A	$pH \rightarrow ml$
R_1	n	a	$pK_a \rightarrow n$
R_2	$\sqrt{I}$	b	$I\uparrow$
R_3	$2n - 1$	c	$T(^\circ C)\uparrow$
R_4	dpK_a/dT	d	$dpK_a/dT\uparrow$
R_5	pK_a'		
R_6	pH		
R_7	$10^{(pH - pK_a')}$		
R_8	I		
R_9	A^{2-}		
R_A	HA^-		
R_B	$T(^\circ C)$		

User Instructions--Algebraic System

Step	Instructions	Input	Keys	Output
1	Enter pK_a	pK_a	2nd A'	
2	Enter number of charges	n	2nd B'	$(2n - 1)$
3	Enter concentration of buffer	Conc.	2nd C'	
4	Temperature entry option default for 25°C	T	B	
5	Enter temperature coefficient for buffer	dpK_a/dT	R/S	

Step	Instructions	Input	Keys	Output
6	Enter pH; calculate number of ml HA^- or HA^{2-} required	pH	A	ml HA^-
7	Calculate number of ml 1 M HA^{2-} required		R/S	HA^{2-}

Program Listing--Algebraic System

Line	Key	Entry	Comments	Line	Key	Entry	Comments
000	76	LBL		030	18	C'	Enter con-
001	16	A'	Enter p$K_a^{25°C}$	031	42	STO	centration of
002	42	STO		032	10	10	buffer
003	01	01		033	34	⌈X	
004	91	R/S		034	42	STO	
005	76	LBL		035	03	03	HA^- option
006	17	B'	Enter number of charges	036	91	R/S	
007	42	STO		037	76	LBL	
008	02	02		038	12	B	Temperature option
009	65	×		039	42	STO	
010	02	2		040	20	20	
011	75	-		041	91	R/S	
012	01	1		042	42	STO	d pK_a/dT
013	95	=		043	05	05	
014	42	STO		044	43	RCL	
015	04	04	$(2n - 1)$	045	01	01	
016	43	RCL		046	85	+	
017	02	02		047	43	RCL	
018	32	X¦T	Put n into t_1	048	05	05	
019	43	RCL	if $x \geq t$	049	65	×	
020	04	04		050	53	(	
021	77	GE		051	43	RCL	
022	38	SIN		052	20	20	
023	91	R/S		053	75	-	
024	76	LBL	If $x \geq t$, set flag	054	02	2	
025	38	SIN		055	05	5	
026	86	STF		056	54	)	
027	01	01		057	95	=	Temperature corrected pK_a
028	91	R/S		058	42	STO	
029	76	LBL		059	01	01	

Line	Key	Entry	Comments
060	91	R/S	
061	76	LBL	Enter pH
062	11	A	
063	42	STO	
064	08	08	
065	53	(	
066	43	RCL	
067	03	03	
068	65	×	
069	93	.	
070	05	5	
071	54	)	
072	55	÷	
073	53	(	
074	53	(	
075	43	RCL	
076	03	03	
077	85	+	
078	01	1	
079	54	)	
080	75	-	
081	53	(	
082	93	.	
083	01	1	
084	65	×	
085	43	RCL	
086	10	10	
087	54	)	
088	95	=	
089	42	STO	
090	05	05	
091	53	(	
092	43	RCL	
093	01	01	
094	75	-	
095	53	(	
096	43	RCL	
097	04	04	
098	65	×	
099	43	RCL	
100	05	05	
101	54	)	
102	95	=	
103	42	STO	

Line	Key	Entry	Comments
104	06	06	pK_a'
105	43	RCL	
106	08	08	
107	75	-	Antilog
108	43	RCL	$(pH - pK_a)$
109	06	06	
110	95	=	
111	22	INV	
112	28	LOG	
113	95	=	
114	42	STO	
115	09	09	
116	87	IFF	
117	01	01	Flag for HA^-
118	39	COS	option
119	43	RCL	
120	03	03	
121	55	÷	
122	43	RCL	
123	09	09	
124	65	×	
125	01	1	
126	00	0	ml of 1 M HA^-
127	00	0	required to
128	00	0	make one liter
129	95	=	of buffer
130	91	R/S	
131	76	LBL	
132	39	COS	
133	43	RCL	
134	10	10	
135	55	÷	
136	53	(	
137	01	1	
138	85	+	
139	53	(	
140	03	3	
141	65	×	
142	43	RCL	
143	09	09	
144	54	)	
145	95	=	
146	42	STO	$xI/(1 + 3x)$
147	13	13	

Line	Key	Entry	Comments	Line	Key	Entry	Comments
148	65	×		166	53	(	
149	01	1	ml of 1M HA^{2-}	167	03	3	
150	00	0	required to	168	65	×	
151	00	0	make one liter	169	43	RCL	
152	00	0	of buffer	170	09	09	
153	95	=		171	54	)	
154	91	R/S		172	95	=	
155	53	(		173	42	STO	
156	43	RCL		174	11	11	
157	09	09		175	65	×	
158	65	×		176	01	1	ml of 1 M HA
159	43	RCL		177	00	0	required to
160	10	10		178	00	0	make one
161	54	)		179	00	0	liter of
162	55	÷		180	95	=	buffer
163	53	(		181	91	R/S	
164	01	1					
165	85	+					

Register Contents, Labels, and Data Cards--Algebraic System

Register	Contents	Labels	Contents
R01	$pK_a^{25°C}$	Label A'	Enter pK_a
R02	Number of charges, n	Label B'	Enter n
R03	$\sqrt{I}$	Label C'	Enter I
R04	$2n - 1$	Label A	Enter pH; calculate
R05	dpK_a/dT	Label B	Enter temperature
R06	pK_a'		
R08	pH		
R09	Antilog (pH - pK_a')		

Register	Contents
R10	Conc. of buffer
R11	A^{2-}
R13	HA^-
R20	Temp. (°C)

Example

Calculate the number of milliliters of each species of phosphate solution required to make a 0.1 M buffer, pH 7.0 at 25°C.
From Appendix III

$$H_2PO_4^-: \quad pK_a = 7.20 \text{ at } 25°C$$
$$I = 0.1$$
$$n = 2$$

Solution

$H_2PO_4^-$: 18.63 ml of 1 M stock solution

HPO_4^{2-}: 27.12 ml of 1 M stock solution plus water to liter of total solution

User Instructions--RPN

Step	Instruction	Input	Keys	Output
1	Enter pK_a	pK_a	ENTER	
2	Enter number of charges	n	A	
3	Enter buffer concentration	I	B	
4	Enter temperature; default is 25°C	T	f a	
5	Enter temperature coefficient of buffer	dpK_a/dT	f b	
6	Enter desired pH and calculate volume (ml) of HCl required for 100 ml solution total volume	pH	C	ml HCl
7	Calculate volume (ml) of base required for 100 ml solution total volume		R/S	ml base

Program Listing--RPN

Line	Key	Comments
001	*LBLA	
002	ST01	Store n (number of
003	2	charges)
004	x	
005	1	
006	-	
007	ST03	$\rightarrow 2n - 1$
008	R↓	
009	ST00	$\rightarrow pK_a^{25°C}$
010	RTN	
011	*LBLB	
012	ST08	$\rightarrow I$
013	√X	
014	ST02	$\rightarrow \sqrt{I}$
015	RTN	
016	*LBLa	
017	ST0A	Store T (°C)
018	RTN	
019	*LBLb	
020	ST04	Optional: Enter
021	RCLA	dpK_a/dT
022	2	
023	-	
024	x	
025	RCL0	
026	+	Temperature corrected
027	ST00	pK_a
028	RTN	
029	*LBLC	
030	ST06	Enter desired pH
031	RCL2	
032	2	
033	÷	
034	RCL2	
035	1	
036	+	
037	.	
038	1	
039	RCL8	
040	x	
041	-	
042	÷	
043	ST04	
044	RCL0	
045	RCL3	

Line	Key	Comments
046	RCL4	
047	x	
048	+	
049	ST05	$\rightarrow pK_a'$
050	RCL6	
051	X⇌Y	
052	-	
053	10ˣ	
054	ST07	$10^{(pH - pK_a')}$
055	RCL8	
056	X⇌Y	
057	1	$I/[1 + (pH - pK_a')]$
058	+	
059	÷	
060	ST09	$\rightarrow C_a$
061	EEX	
062	3	
063	x	
064	R/S	$\rightarrow$ ml HCl required
065	RCL8	
066	RCL9	
067	-	
068	EEX	
069	3	
070	x	
071	RTN	$\rightarrow$ ml base required

Register Contents, Labels, and Data Cards--RPN

Register	Contents	Labels	Contents
R_0	pK_a	A	$pK_a \uparrow n$
R_1	n	B	$I \uparrow$
R_2	$\sqrt{I}$	C	pH $\rightarrow$ ml HCl; ml base
R_3	$2n - 1$	a	$T(°C) \uparrow$
R_4	dpK_a/dT	b	dpK_a/dT
R_5	pK_a'		
R_6	pH		
R_7	$10^{(pH - pK_a')}$		
R_8	I		
R_9	c_a		
R_A	$T(°C)$		

User Instructions--Algebraic System

Step	Instructions	Input	Keys	Output
1	Enter pK_a	pK_a	2nd A'	
2	Enter number of charges	n	2nd B'	
3	Enter concentration of buffer	Conc.	2nd C'	
4	Temperature entry option default for 25°C	T	B	

Step	Instructions	Input	Keys	Output
5	Enter temperature coefficient for buffer	dpK_a/dT	R/S	
6	Enter pH; calculate ml HCl required for 100 ml solution total volume	pH	A	ml HCl
7	Calculate ml base required for 100 ml solution total volume		R/S	ml base

Program Listing--Algebraic System

Line	Key	Entry	Comments	Line	Key	Entry	Comments
000	76	LBL		028	42	STO	
001	16	A'	Enter pK_a	029	20	20	Enter dpK_a/dT
002	42	STO		030	91	R/S	
003	01	01		031	42	STO	
004	91	R/S		032	05	05	
005	76	LBL		033	53	(	
006	17	B'	Enter n	034	43	RCL	
007	42	STO		035	01	01	
008	02	02		036	85	+	
009	65	×		037	53	(	
010	02	2		038	43	RCL	
011	54	)		039	05	05	
012	75	-		040	65	×	
013	01	1		041	53	(	
014	95	=		042	43	RCL	
015	42	STO		043	20	20	
016	04	04		044	75	-	
017	91	R/S		045	02	2	
018	76	LBL	Enter I	046	05	5	
019	18	C'		047	54	)	
020	42	STO		048	95	=	
021	10	10		049	42	STO	Enter pH
022	34	ГX	$\sqrt{I}$	050	01	01	
023	42	STO		051	91	R/S	
024	03	03		052	76	LBL	
025	91	R/S	Temperature	053	11	A	
026	76	LBL	option	054	42	STO	
027	12	B		055	08	08	

Line	Key	Entry	Comments	Line	Key	Entry	Comments
056	53	(		099	08	08	
057	53	(		100	75	-	
058	53	(		101	43	RCL	antilog
059	43	RCL		102	06	06	$(pH - pK_a)$
060	03	03		103	95	=	
061	65	×		104	22	INV	
062	93	.		105	28	LOG	
063	05	5		106	95	=	
064	54	)		107	42	STO	
065	55	÷		108	09	09	
066	53	(		109	43	RCL	
067	53	(		110	10	10	$I/[1 + (pH -$
068	43	RCL		111	55	÷	$pK_a)]$
069	03	03		112	53	(	
070	85	+		113	01	1	
071	01	1		114	85	+	
072	54	)		115	43	RCL	c_a
073	75	-		116	09	09	
074	53	(		117	54	)	
075	93	.		118	95	=	
076	01	1		119	42	STO	
077	65	×		120	15	15	ml HCl
078	43	RCL		121	65	×	required
079	10	10		122	01	1	
080	54	)		123	00	0	
081	95	=		124	00	0	
082	42	STO		125	00	0	
083	05	05		126	95	=	
084	53	(		127	91	R/S	
085	43	RCL		128	53	(	
086	01	01		129	43	RCL	
087	85	+		130	10	10	
088	53	(		131	75	-	
089	43	RCL		132	43	RCL	
090	04	04		133	15	15	ml base
091	65	×	pK_a'	134	54	)	required
092	43	RCL		135	65	×	
093	05	05		136	01	1	
094	54	)		137	00	0	
095	95	=		138	00	0	
096	42	STO		139	00	0	
097	06	06		140	95	=	
098	43	RCL		141	91	R/S	

Register Contents, Labels, and Data Cards--Algebraic System

Register	Contents	Labels	Contents
R01	pK_a	Label A'	Enter pK_a
R02	n	Label B'	Enter n
R03	$\sqrt{I}$	Label C'	Enter I
R04	$2n - 1$	Label A	Enter pH; calculate
R05	dpK_a/dT	Label B	Enter temperature
R06	pK_a'		
R08	pH		
R09	antilog (pH $-$ pK_a')		
R10	I		
R15	c_a		
R20	Temperature (°C)		

Example

Calculate the number of milliliters of HCl required to make a 0.1 M N-ethylmorpholine buffer pH 7.8 at 25°C.
From Appendix III

$$N\text{-ethylmorpholine:} \quad pK_a = 7.67 \text{ at } 25°C$$
$$I = 0.1$$
$$n = 1$$

Solution

HCl: 0.1 M = 49.5 ml
N-ethylmorpholine: 0.2 M = 50.5 ml
Dilute to 100 ml with water to make
0.1 M buffer solution.

References

1. D. Perrin and B. Dempsey (1974), *Buffers for pH and Metal Ion Control*, John Wiley & Sons, Inc., New York, pp. 62-76.

1B. TITRATION OF A WEAK ACID OR WEAK BASE

This section and the accompanying calculator program are derived from a program contributed by Karl Marhenke to the HP-67/ 97 User's Library. Hewlett-Packard subsequently published this program in the *User's Library Solutions in Chemistry*.

Consider a weak acid, H_4A. The electroneutrality of any solution containing this acid and its ions requires that

$$[H^+] + [Na^+] = [OH^-] + [H_3A^-] + 2[H_2A^{2-}]$$
$$+ 3[HA^{3-}] + 4[A^{4-}] \tag{1}$$

The Na^+ term must be included once neutralization is begun; NaOH is assumed to be the titrant. The acid ion concentrations must be expressed in terms of the dissociation constants and of C, the "analytical concentration" of the acid:

$$C = \frac{V_a M_a}{V_a + V_b}$$

where V_a = initial volume of H_4A

M_a = molar concentration of H_4A

V_b = volume of titrant added

See any beginning text on quantitative analysis for the derivation of the formulas for the fraction of each ionic species as a function of H^+. Let

$$Q = 1 + \frac{K_1}{[H^+]} + \frac{K_1 K_2}{[H^+]^2} + \frac{K_1 K_2 K_3}{[H^+]^3} + \frac{K_1 K_2 K_3 K_4}{[H^+]^4} \tag{2}$$

where K_i = the equilibrium constant of the i'th dissociation step of a polyprotic acid. For each acid ion concentration in equation 1, we now can substitute its fraction times C:

$$H^+ + Na^+ = \frac{K_w}{[H^+]} + \frac{CK_1}{Q[H^+]} + \frac{2CK_1 K_2}{Q[H^+]^2} + \frac{3CK_1 K_2 K_3}{Q[H^+]^3}$$
$$+ \frac{4CK_1 K_2 K_3 K_4}{Q[H^+]^4}$$

where K_w = dissociation constant of water. After substituting for Q, clearing fractions, and collecting terms, the following sixth degree equation is derived:

$$[H^+]^6 + (K_1 + [Na^+])[H^+]^5 + (K_1K_2 + [Na^+]K_1 - CK_1 - K_w)[H^+]^4$$

$$+ (K_1K_2K_3 + [Na^+]K_1K_2 - 2CK_1K_2 - K_1K_w)[H^+]^3$$

$$+ (K_1K_2K_3K_4 + [Na^+]K_1K_2K_3 - 3CK_1K_2K_3 - K_1K_2K_w)[H^+]^2$$

$$+ ([Na^+]K_1K_2K_3K_4 - 4CK_1K_2K_3K_4 - K_1K_2K_3K_w)[H^+]$$

$$- K_1K_2K_3K_4K_w = 0$$

Using letters a, b, c, d, e, and f to represent the coefficients gives

$$[H^+]^6 + a[H^+]^5 + b[H^+]^4 + c[H^+]^3 + d[H^+]^2 + e[H^+] + f = 0$$

The solution of this sixth degree polynomial requires a Newton-Raphson method for finding roots. The Newton-Raphson method for finding roots of functions is discussed in most beginning calculus texts. If the polynomial in hydrogen ion on the left side of the above equation is called $g[H^+]$, the Newton-Raphson formula is

$$[H^+]n+1 = [H^+]n - \left(\frac{g[H^+]n}{(g'[H^+]n)}\right)$$

where $[H^+]n$ = a trial value of hydrogen ion concentration

$[H^+]n+1$ = a new value of $[H^+]$, closer to the root of the equation than $[H^+]_n$

$g[H^+]n$ = the function g evaluated at $[H^+] = [H^+]_n$

$g'[H^+]n$ = the first derivative of g evaluated at $[H^+] = [H^+]_n$

Here

$$g'[H^+] = 6[H^+]^5 + 5a[H^+]^4 + 4b[H^+]^3 + 3c[H^+]^2 + 2d[H^+] + e$$

The iteration process continues, using each value of $[H^+]$ generated as the trial value for the next iteration, until a value for hydrogen ion is generated which differs by 1% or less from the preceding value. The last value of hydrogen ion obtained is then converted to pH and presented as the answer. The 1% figure corresponds to 0.0043 in the pH, which means that these calculated pH's should agree about as well as can be expected with values obtained in the laboratory, since junction potentials, activity coefficients, and so on are not taken into account.

In every case checked by Karl Marhenke, all six (or five, four, or three) roots are real, but only one (the one of interest) is positive. If the initial pH estimate (pH_{est}) corresponding to too small a [H^+] is taken, it is quite possible for the program to iterate its way to the largest of the negative roots. To prevent an "error" message being produced when the calculator tries to take the log of a negative root, each value of hydrogen ion produced is checked at step 119 to see whether it is negative. If it is, the program returns to Label a (step 080), the pH_{est} is lowered (or raised, if the calculator is in "base mode") by two units, and the iteration process starts anew. The calculator will thus always converge to the positive root, but the process can be quite lengthy if a poor initial guess is made.

If a tribasic acid, H_3A, is used, K_4 is zero and the sixth degree equation given reduces to the fifth degree equation:

$$[H^+]^5 + (K_1 + [Na^+])[H^+]^4 + (K_1K_2 + [Na^+]K_1 - CK_1 - K_w)[H^+]^3$$

$$+ (K_1K_2K_3 + [Na^+]K_1K_2 - 2CK_1K_2 - K_1K_w)[H^+]^2$$

$$+ [Na^+]K_1K_2K_3 - 3CK_1K_2K_3 - K_1K_2K_3)[H^+] - K_1K_2K_3K_w = 0$$

Here $g[H^+] = [H^+]^5 + a[H^+]^4 + b[H^+]^3 + c[H^+]^2 + d[H^+] + e$

and $g'[H^+] = 5[H^+]^4 + 4a[H^+]^3 + 3b[H^+]^2 + 2c[H^+] + d$

If a dibasic acid, H_2A, is used, K_3 and K_4 are both zero, and the sixth degree equation reduces to the fourth degree equation:

$$[H^+]^4 + (K_1 + [Na^+])[H^+]^3 + (K_1K_2 + [Na^+]K_1 - CK_1 - K_w)[H^+]^2$$

$$+ ([Na^+]K_1K_2 - 2CK_1K_2 - K_1K_w)[H^+] - K_1K_2K_w = 0$$

Here $g[H^+] = [H^+]^4 + a[H^+]^3 + b[H^+]^2 + c[H^+] + d$

and $g'[H^+] = 4[H^+] + 3a[H^+]^2 + 2b[H^+] + c$

Finally, if a monobasic acid, HA, is used, K_2, K_3, and K_4 are all zero, and the sixth degree equation reduces to

$$[H^+]^3 + (K + [Na^+])[H^+]^2 + ([Na^+]K - CK - K_w)[H^+] - KK_w = 0$$

Here $g[H^+] = [H^+]^3 + a[H^+]^2 + b[H^+] + c$

and $g'[H^+] = 3[H^+]^2 + 2a[H^+] + b$

If the weak electrolyte being titrated is a base rather than an acid, the mathematical treatment is identical. However, the equation that must be solved is an equation in $[OH^-]$ rather than in $[H^+]$, and the role of $[Na^+]$ is assumed by $[Cl^-]$, assuming HCl as the titrant. Subroutine Label d converts pH to pOH or vice versa, since even when titrating a base, the answer is pH and not pOH.

If the volume of titrant added is zero, the program automatically takes $\sqrt{K_1 C}$ as its first trial $[H^+]$. For titrant volumes greater than zero, usually the most practical trial pH to use is the one that was obtained for the preceding volume (assuming that the preceding volume was smaller than the present one). It is not even necessary to key this value in, since it is already in the display.

User Instructions--RPN

Step	Instructions	Input	Keys	Output
1	Initialize: Clear registers		f e	0.00
2	Optional: Set "base mode"		C	
	Clear "base mode"		C	1.00
3.	Enter dissociation constants	K_1	E	K_1
		K_2	E	K_2
		K_3	E	K_3
		K_4	E	K_4
4.	Enter: a. volume of weak acid or base to be titrated	V_a	ENTER↑	
	b. molar concentration of strong base or acid to be titrated	M_a	ENTER↑	
	c. normality of strong base or acid titrant	N	D	Number of K's
5.	Semioptional: Enter guess for pH*	pH_{est}	ENTER↑	
	Enter volume of titrant added	V_t	A	pH
6.	Repeat step 5 as often as needed to plot curve			

*A guess for pH need <u>not</u> be entered if one of the following is true:
1. The volume of titrant is to be zero.
2. The pH already in the display (usually from the problem just done) is a suitable pH_{est}; it usually is suitable if the volume to be entered is greater than the volume just used.

Program Listing--RPN

Line	Key	Comments	Line	Key	Comments
001	*LBLC		046	4	
002	0	Acid mode--Set	047	-	
003	F0?	Base mode--Clear	048	*LBLb	
004	1		049	DSZI	
005	SF0		050	GTO0	
006	X≠0?		051	RCL7	
007	CF0		052	X≷I	
008	RTN		053	X≷Y	
009	*LBLA		054	P≷S	
010	X=0?	Volume = 0?	055	STOi	
011	SF2	Yes: set flag 2	056	P≷S	
012	PRTX	No: flag 2 clear	057	7	
013	RCLB		058	STOI	
014	X≷Y		059	DSZi	
015	x		060	GTOc	
016	LSTX		061	RCL0	→ Number of K's
017	RCLD		062	STOI	R_I now ready for
018	+		063	RCL1	step 094
019	÷		064	RCLA	
020	STOA	→ $[Na^+]$	065	+	
021	RCLE		066	P≷S	
022	LSTX		067	STO0	Store a in R_{s0}
023	÷		068	P≷S	
024	STOC	→ C	069	RCL1	→ K_1
025	R↑		070	RCLC	→ C
026	STO8	→ pH_{est}	071	x	
027	RCL0		072	√X	$\sqrt{K_1 C}$ = $[H^+]$ trial if
028	STO7		073	F2?	Volume = 0
029	*LBLc		074	GTO1	Volume = 0?
030	RCL7	Initialize	075	GTOB	Yes: begin
031	STOI	coefficient counter	076	*LBL0	iterations
032	ISZI		077	RCLi	No: use pH_{est}
033	RCLi		078	x	first
034	RCLA		079	GTOb	
035	+		080	*LBLa	Change pH_{est} by ± 2
036	DSZI		081	2	
037	RCLI	Calculate	082	F0?	
038	RCLC	coefficients	083	CHS	
039	x		084	ST+8	
040	-		085	*LBLB	
041	RCLi		086	RCL8	
042	x		087	GSBd	
043	EEX		088	CHS	
044	CHS		089	10^x	$[H^+]$ = 10^{-pH}
045	1		090	*LBL1	

Line	Key	Comments
091	P≵S	
092	ENT↑	Load stack with
093	ENT↑	$[H^+]$
094	GTOi	→ Use correct N-R
095	*LBL5	routine
096	6	Routine for four
097	x	K's:
098	RCL0	
099	5	
100	GSB6	Start forming
101	RCL1	g' $[H^+]$
102	4	
103	GSB6	
104	RCL2	
105	3	
106	GSB6	
107	RCL3	
108	2	
109	GSB6	
110	RCL4	
111	GSB8	
112	RCL3	
113	GSB7	
114	RCL4	
115	GSB7	
116	RCL5	
117	*LBL9	Finish forming
118	+	g' $[H^+]$
119	RCL6	RCL g' $[H^+]$ forms
120	P≵S	$[H^+]$
121	÷	
122	-	
123	X<0?	[H] < 0?
124	GTOa	Yes: start over
125	STO9	with new pH
126	%CH	No: continue
127	ABS	
128	1	
129	X≤Y?	1% ≤ |%CH|?
130	SF2	Yes: set flag 2
131	RCL9	No: flag 2 clear
132	F2?	1% ≤ |%CH|?
133	GTO1	Yes: another
134	LOG	iteration
135	CHS	No: $[H^+]$ → pH
136	GSBd	
137	PRTX	

Line	Key	Comments
138	SPC	
139	RTN	
140	*LBLd	
141	F0?	Acid mode:
142	RTN	Yes: leave pH alone
143	1	No: 14 - pH = pOH
144	4	or 14 - pOH = pH
145	X≵Y	
146	-	
147	RTN	
148	*LBL6	Repeatedly used in
149	x	polynomial evalua-
150	*LBL7	tion
151	+	Steps common to all
152	x	N-R routines:
153	RTN	
154	*LBL8	
155	+	
156	STO6	
157	CLX	Completes and stores
158	RCL0	g' $[H^+]$ in R_6; begins
159	GSB7	computation of $g[H^+]$
160	RCL1	
161	GSB7	
162	RCL2	
163	F2?	Flag 2 set in LBL 2
164	GTO9	only
165	GTO7	
166	*LBL4	Routine for three
167	5	K's:
168	x	
169	RCL0	Start forming $g'[H^+]$
170	4	
171	GSB6	
172	RCL1	
173	3	
174	GSB6	
175	RCL2	
176	2	
177	GSB6	
178	RCL3	
179	GSB8	
180	RCL3	
181	GSB7	
182	RCL4	
183	GTO9	
184	*LBL3	

Line	Key	Comments
185	4	Routine for two
186	x	K'
187	RCL0	Start forming $g'[H^+]$
188	3	
189	GSB6	
190	RCL1	
191	2	
192	GSB6	
193	RCL2	
194	GSB8	
195	RCL3	
196	GTO9	
197	*LBL2	Routine for one K:
198	3	Start forming $g'[H^+]$
199	x	
200	RCL0	
201	2	
202	GSB6	
203	RCL1	
204	SF2	
205	GTO8	Cause LBL 8 to exit
206	*LBLe	to LBL 9
207	CLRG	Initialize R_0 to 1
208	ISZi	
209	RTN	
210	*LBLE	
211	ISZI	$K_i \rightarrow R_i$
212	STOi	
213	RTN	
214	*LBLD	
215	STOB	$N_b \rightarrow R_B$
216	R↓	
217	X⇄Y	
218	STOD	$V_a \rightarrow R_D$
219	x	
220	STOE	$V_a M_a \rightarrow R_E$
221	RCLI	Number of K's
222	ST+0	Number of
223	RTN	K's $+ 1 \rightarrow R_0$

Register Contents, Labels, and Data Cards--RPN

Register	Contents	Labels	Contents
R_0	Number of K's + 1	A	$pH_{est} \uparrow V_a \to pH$
R_1	K_1	C	Base?
R_2	K_2	D	$V_a \uparrow M_a \to N$
R_3	K_3	E	$K_1; K_2; K_3; K_4$
R_4	K_4	e	Initialize
R_7	Counter		
R_8	pH_{est}		
R_9	$[H^+]_{trial}$		
R_{s0}	a		
R_{s1}	b		
R_{s2}	c		
R_{s3}	d		
R_{s4}	e		
R_{s5}	f		
R_{s6}	$g'[H^+]$		
R_A	$[Na^+]$		
R_B	N		
R_C	C		
R_D	V_b		
R_E	$V_a M_a$		
R_I	Control		

User Instructions--Algebraic System

Step	Instruction	Input	Keys	Output
1	Initialize: Clear memories and flags		2nd E'	0.00
2	Set "acid mode: Clear "acid mode"		C C	1.00 0.00
3	Enter dissociation constants	K_1 K_2 K_3 K_4	E E E E	
4	Enter volume of weak acid or base to be titrated	V_a	D	
5	Enter molar concentration of weak acid or base to be titrated	M_a	R/S	
6	Enter normality of strong base or acid titrant	N	R/S	Number of K's
	(Semi-optional): Enter guess for pH*	pH_{est}	A	
7	Enter volume of titrant added	V_t	B	pH

Repeat step 7 as often as needed
to plot curve

*A guess for pH need <u>not</u> be entered if:
   1. The value of titrant is to be zero, or
   2. The pH already in the display (usually from the problem just done) is a suitable pH_{est}; it usually is suitable if the volume to be entered is greater than the value just used.

Program Listing--Algebraic System

Line	Key	Entry	Comments	Line	Key	Entry	Comments
000	76	LBL		042	42	STO	
001	10	E'		043	12	12	$M_a \to R_{12}$
002	47	CMS	Initialize	044	91	R/S	
003	69	OP	R_0 to 1	045	42	STO	$N_b \to R_{13}$
004	20	20	R_6 to 20	046	13	13	
005	02	2		047	43	RCL	
006	00	0		048	12	12	
007	42	STO		049	65	×	
008	06	06		050	43	RCL	
009	22	INV	Clear flags	051	11	11	
010	86	STF		052	95	=	
011	00	00		053	42	STO	$V_a M_a \to R_{17}$
012	22	INV		054	17	17	
013	86	STF		055	43	RCL	
014	01	01		056	09	09	
015	22	INV		057	44	SUM	
016	86	STF		058	00	00	
017	02	02		059	92	RTN	
018	22	INV		060	76	LBL	
019	86	STF		061	11	A	
020	03	03		062	42	STO	$pH_{est} \to R_8$
021	22	INV		063	08	08	
022	86	STF		064	92	RTN	
023	04	04		065	76	LBL	
024	25	CLR		066	12	B	
025	58	FIX		067	42	STO	$V_b \to R_{15}$
026	03	03		068	15	15	
027	92	RTN		069	00	0	
028	76	LBL		070	32	X⇄T	
029	15	E		071	43	RCL	
030	69	OP		072	15	15	
031	29	29	$K_i \to R_i$	073	67	EQ	Volume = 0 ?
032	72	ST*		074	39	COS	Yes: set
033	09	09		075	76	LBL	flag 2
034	92	RTN		076	38	SIN	No: print
035	76	LBL		077	99	PRT	and continue
036	14	D		078	43	RCL	
037	22	INV		079	15	15	
038	52	EE		080	65	×	
039	42	STO	$V_a \to R_{11}$	081	43	RCL	
040	11	11		082	13	13	
041	91	R/S		083	95	=	

Line	Key	Entry	Comments	Line	Key	Entry	Comments
084	55	÷		128	05	05	
085	53	(		129	00	0	
086	43	RCL		130	69	OP	
087	15	15		131	39	39	
088	85	+		132	53	(	
089	43	RCL		133	43	RCL	
090	11	11		134	05	05	
091	95	=		135	75	-	
092	42	STO	$[Na^+] \rightarrow R_{16}$	136	53	(	
093	16	16		137	43	RCL	
094	43	RCL		138	09	09	
095	17	17		139	65	×	
096	55	÷		140	43	RCL	
097	53	(		141	14	14	
098	43	RCL		142	54	)	
099	15	15		143	54	)	
100	85	+		144	65	×	
101	43	RCL		145	73	RC*	
102	11	11		146	09	09	
103	95	=		147	95	=	
104	42	STO	$C \rightarrow R_{14}$	148	75	-	
105	14	14		149	01	1	
106	43	RCL		150	52	EE	
107	00	00		151	01	1	
108	42	STO		152	04	4	
109	07	07		153	94	+/-	
110	44	SUM		154	95	=	
111	06	06		155	76	LBL	
112	76	LBL	Initialize	156	17	B'	
113	18	C'	coefficient	157	97	DSZ	
114	43	RCL	counter	158	09	09	
115	07	07		159	95	=	
116	42	STO		160	72	ST*	
117	09	09		161	06	06	
118	69	OP		162	69	OP	
119	29	29	Calculate	163	36	36	
120	53	(	coefficients	164	97	DSZ	
121	73	RC*		165	07	07	
122	09	09		166	18	C'	
123	85	+		167	43	RCL	Number of K's
124	43	RCL		168	00	00	R_9 now ready
125	16	16		169	42	STO	for iterat-
126	95	=		170	09	09	ions
127	42	STO		171	43	RCL	

Line	Key	Entry	Comments	Line	Key	Entry	Comments
172	01	01		216	22	INV	
173	85	+		217	86	STF	Determine
174	43	RCL		218	02	02	which iterat-
175	16	16		219	43	RCL	ion routine
176	95	=		220	00	00	to use
177	42	STD	Store $a \to R_{20}$	221	32	X:T	
178	20	20		222	05	5	
179	43	RCL		223	67	EQ	
180	01	01		224	58	FIX	
181	65	×		225	04	4	
182	43	RCL		226	67	EQ	
183	14	14		227	60	DEG	
184	95	=		228	03	3	
185	34	ΓΧ		229	67	EQ	
186	42	STD	$\sqrt{K_1 C} \to R_{10}$	230	80	GRD	
187	10	10		231	02	2	
188	87	IFF	Volume = 0 ?	232	67	EQ	
189	02	02	Yes: begin	233	70	RAD	
190	61	GTD	iterations	234	91	R/S	
191	61	GTD	No: use pH_{est}	235	76	LBL	
192	35	1/X	first	236	58	FIX	Routine for
193	76	LBL		237	43	RCL	four K's:
194	16	A'		238	10	10	
195	02	2	Change pH by	239	45	Y×	
196	87	IFF	± 2	240	05	5	
197	00	00		241	65	×	
198	30	TAN		242	06	6	
199	76	LBL		243	85	+	
200	89	π		244	43	RCL	
201	44	SUM		245	10	10	
202	08	08		246	45	Y×	
203	76	LBL		247	04	4	
204	35	1/X		248	65	×	
205	43	RCL		249	43	RCL	
206	08	08		250	20	20	
207	71	SBR		251	65	×	
208	19	D'		252	05	5	
209	94	+/-		253	85	+	
210	22	INV		254	43	RCL	
211	28	LOG		255	10	10	
212	42	STD		256	45	Y×	
213	10	10		257	03	3	
214	76	LBL		258	65	×	
215	61	GTD		259	43	RCL	

Line	Key	Entry	Comments	Line	Key	Entry	Comments
260	21	21		304	43	RCL	
261	65	×		305	21	21	
262	04	4		306	85	+	
263	85	+		307	43	RCL	
264	43	RCL		308	10	10	
265	10	10		309	45	Y^x	
266	33	X^2		310	03	3	
267	65	×		311	65	×	
268	43	RCL		312	43	RCL	
269	22	22		313	22	22	
270	65	×		314	85	+	
271	03	3		315	43	RCL	
272	85	+		316	10	10	
273	43	RCL		317	33	X^2	
274	10	10		318	65	×	
275	65	×		319	43	RCL	
276	43	RCL		320	23	23	
277	23	23		321	85	+	
278	65	×		322	43	RCL	
279	02	2		323	10	10	
280	85	+		324	65	×	
281	43	RCL		325	43	RCL	
282	24	24		326	24	24	
283	95	=		327	85	+	
284	42	STO	$g'[H^+] \to R_{27}$	328	43	RCL	
285	27	27		329	25	25	
286	43	RCL		330	95	=	
287	10	10		331	42	STO	$g[H^+] \to R_{28}$
288	45	Y^x		332	28	28	
289	06	6		333	76	LBL	
290	85	+		334	52	EE	Compute
291	43	RCL		335	43	RCL	$[H^+]_{n+1}$
292	10	10		336	10	10	
293	45	Y^x		337	75	-	
294	05	5		338	43	RCL	
295	65	×		339	28	28	
296	43	RCL		340	55	÷	
297	20	20		341	43	RCL	
298	85	+		342	27	27	
299	43	RCL		343	95	=	
300	10	10		344	42	STO	
301	45	Y^x		345	29	29	
302	04	4		346	00	0	
303	65	×		347	32	X⁀T	

Line	Key	Entry	Comments	Line	Key	Entry	Comments
348	43	RCL		392	08	08	
349	29	29		393	99	PRT	Print pH
350	22	INV	$[H^+]_{n+1} \leq 0$?	394	98	ADV	
351	77	GE		395	92	RTN	
352	16	A'	Yes: start	396	76	LBL	
353	43	RCL	over with new	397	28	LOG	
354	10	10	pH_{est}	398	86	STF	
355	75	-	No: continue	399	02	02	
356	43	RCL		400	61	GTO	
357	29	29		401	23	LNX	
358	95	=		402	76	LBL	
359	55	÷		403	95	=	
360	43	RCL		404	65	×	
361	10	10		405	73	RC*	
362	95	=		406	09	09	
363	65	×		407	95	=	
364	01	1		408	42	STO	
365	00	0		409	19	19	
366	00	0		410	61	GTO	
367	95	=		411	17	B'	
368	50	I×I	$[H^+]_{n+1} \geq$	412	76	LBL	
369	32	X:T	$[H^+]_n$ by 1% ?	413	30	TAN	
370	01	1		414	94	+/-	
371	77	GE	Yes: another	415	61	GTO	
372	28	LOG	iteration	416	89	π	
373	76	LBL	No: convert	417	76	LBL	
374	23	LNX	$[H^+]$ to pH	418	19	D'	Acid mode?
375	43	RCL		419	87	IFF	Yes: leave
376	29	29		420	00	00	pH alone
377	42	STO		421	34	ГX	No: change
378	10	10		422	75	-	pH to pOH or
379	22	INV		423	01	1	vice versa
380	87	IFF		424	04	4	
381	02	02		425	95	=	
382	61	GTO		426	94	+/-	
383	28	LOG		427	92	RTN	
384	94	+/-		428	76	LBL	
385	71	SBR		429	34	ГX	
386	19	D'		430	92	RTN	
387	22	INV		431	76	LBL	
388	52	EE		432	60	DEG	Routine for
389	58	FIX		433	43	RCL	three K's:
390	03	03		434	10	10	
391	42	STO		435	45	YX	

Line	Key	Entry	Comments	Line	Key	Entry	Comments
436	04	4		480	04	4	
437	65	×		481	65	×	
438	05	5		482	43	RCL	
439	85	+		483	20	20	
440	43	RCL		484	85	+	
441	10	10		485	43	RCL	
442	45	Y×		486	10	10	
443	03	3		487	45	Y×	
444	65	×		488	03	3	
445	43	RCL		489	65	×	
446	20	20		490	43	RCL	
447	65	×		491	21	21	
448	04	4		492	85	+	
449	85	+		493	43	RCL	
450	43	RCL		494	10	10	
451	10	10		495	33	X²	
452	33	X²		496	65	×	
453	65	×		497	43	RCL	
454	43	RCL		498	22	22	
455	21	21		499	85	+	
456	65	×		500	43	RCL	
457	03	3		501	10	10	
458	85	+		502	65	×	
459	43	RCL		503	43	RCL	
460	10	10		504	23	23	
461	65	×		505	85	+	
462	43	RCL		506	43	RCL	
463	22	22		507	24	24	
464	65	×		508	95	=	
465	02	2		509	42	STO	
466	85	+		510	28	28	
467	43	RCL		511	61	GTO	
468	23	23		512	52	EE	
469	95	=		513	76	LBL	
470	42	STO		514	80	GRD	Routine for
471	27	27		515	43	RCL	two K's:
472	43	RCL		516	10	10	
473	10	10		517	45	Y×	
474	45	Y×		518	03	3	
475	05	5		519	65	×	
476	85	+		520	04	4	
477	43	RCL		521	85	+	
478	10	10		522	43	RCL	
479	45	Y×		523	10	10	

Line	Key	Entry	Comments	Line	Key	Entry	Comments
524	33	X²		568	22	22	
525	65	×		569	85	+	
526	43	RCL		570	43	RCL	
527	20	20		571	23	23	
528	65	×		572	95	=	
529	03	3		573	42	STO	
530	85	+		574	28	28	
531	43	RCL		575	61	GTO	
532	10	10		576	52	EE	
533	65	×		577	76	LBL	
534	43	RCL		578	70	RAD	Routine for
535	21	21		579	43	RCL	one K:
536	65	×		580	10	10	
537	02	2		581	33	X²	
538	85	+		582	65	×	
539	43	RCL		583	03	3	
540	22	22		584	85	+	
541	95	=		585	43	RCL	
542	42	STO		586	10	10	
543	27	27		587	65	×	
544	43	RCL		588	43	RCL	
545	10	10		589	20	20	
546	45	Yˣ		590	65	×	
547	04	4		591	02	2	
548	85	+		592	85	+	
549	43	RCL		593	43	RCL	
550	10	10		594	21	21	
551	45	Yˣ		595	95	=	
552	03	3		596	42	STO	
553	65	×		597	27	27	
554	43	RCL		598	43	RCL	
555	20	20		599	10	10	
556	85	+		600	45	Yˣ	
557	43	RCL		601	03	3	
558	10	10		602	85	+	
559	33	X²		603	43	RCL	
560	65	×		604	10	10	
561	43	RCL		605	33	X²	
562	21	21		606	65	×	
563	85	+		607	43	RCL	
564	43	RCL		608	20	20	
565	10	10		609	85	+	
566	65	×		610	43	RCL	
567	43	RCL		611	10	10	

Line.	Key	Entry	Comments
612	65	×	
613	43	RCL	
614	21	21	
615	85	+	
616	43	RCL	
617	22	22	
618	95	=	
619	42	STO	
620	28	28	
621	61	GTO	
622	52	EE	
623	76	LBL	
624	39	COS	Set flag 2 if
625	86	STF	volume = 0
626	02	02	
627	61	GTO	
628	38	SIN	
629	76	LBL	
630	13	C	Set flag 0 for
631	00	0	acid mode
632	87	IFF	
633	00	00	
634	57	ENG	
635	01	1	
636	86	STF	
637	00	00	Clear flag 0
638	92	RTN	for base mode
639	76	LBL	
640	57	ENG	
641	22	INV	
642	86	STF	
643	00	00	
644	92	RTN	

Register Contents, Labels, and Data Cards--ALGEBRAIC SYSTEM

Register	Contents
R_0	number of K's + 1
R_1	K_1
R_2	K_2
R_3	K_3
R_4	K_4
R_6	Index = 20
R_7	Counter
R_8	pH_{est}
R_9	Index for $K_i \rightarrow R_i$
R_{10}	$\sqrt{K_1 C}$
R_{11}	Volume of acid (V_a)
R_{12}	Molarity of acid (M_a)
R_{13}	Normality of base titrant (N_b)
R_{14}	Concentration (C)
R_{15}	Volume of base titrant (V_t)
R_{16}	$[Na^+]$
R_{17}	$V_a M_a$
R_{20}	a
R_{21}	b
R_{22}	c
R_{23}	d

Register	Contents
R_{24}	e
R_{25}	f
R_{27}	$g'([H^+])$
R_{28}	$g([H^+])$
R_{29}	$[H^+]_{n+1}$

Labels (User-defined)	Contents
A	Enter pH_{est}
B	Enter volume of base
C	Base ?
D	$V_a; M_a; N_b$
E	$K_1; K_2; K_3; K_4$

Labels (Common)	Contents
C'	If flag 2 is set--then clear it
COS	If not set $\rightarrow$ then set it (display 1)
SIN	Print volume and continue
GTO	Determine which iteration routine to use (i.e., for 5, 4, 3 or 2nd degree equation)
$1/X$	Use pH_{est} for first approximation
A'	If base mode set, subtract 2 from pH_{est}
π	Otherwise add 2 to pH_{est}

Labels (Common)	Contents
FIX	Routine for four K's
DEG	Routine for three K's
GRD	Routine for two K's
RAD	Routine for one K
EE	Compute $[H^+]_{n+1}$
LNX	$[H^+]_{n+1}$ becomes $[H^+]_n$
D'	If acid mode leave pH alone
$\sqrt{X}$	Return
LOG	Set flag 4 if 1% $\leq$ $\lvert$1% CH$\rvert$
=	Same as LBL 0 of HP program
TAN	change sign
ENG	Clear flag 0

Example

For phosphoric acid, H_3PO_4: $K_1 = 7.5 \times 10^{-3}$

$K_2 = 6.2 \times 10^{-8}$

$K_3 = 1 \quad \times 10^{-12}$

Plot a titration curve from 0 to 75 ml of base added for 50.00 ml of 0.200 M H_2PO_4 titrated with 0.500 N NaOH.

Solution

For a complete solution for purposes of a plot, calculations of the pH at approximately 45 different titrant volumes are required. For purposes of illustration the following 15 calculations will suffice:

V_t^*	pH
0	1.454
10	2.192
15	2.637
19.4	3.654
20	4.677
20.4	5.527
21	5.930
25	6.730
35	7.685
39	8.483
39.8	9.142
40.1	9.832
40.8	10.584
45	11.457
75	12.845

These data are plotted as shown in Figure 1B below.

$^*V_t$ = total volume of titrant added

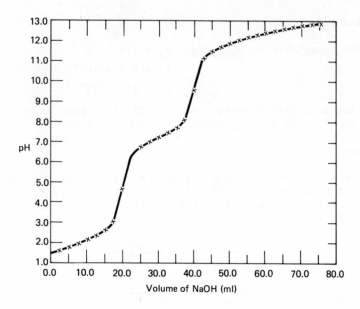

Figure 1B.   Calculated titration curve from example.

REFERENCES

1. G.L. Breneman (1974), "A General Acid-Base Titration Curve
    Computer Program," *Journal of Chemical Education 51*, 812-813.

2. Douglas A. Skoog and Donald M. West (1965), *Analytical
    Chemistry--An Introduction*, Chapter 13.   Holt, Rinehart and
    Winston, New York.

1C. ACID-BASE EQUILIBRIUM

It is often necessary to calculate the hydrogen ion concentration of a solution of a weak acid or base in the preparation of buffers or other solutions in the laboratory. Various textbooks of analytical chemistry provide methods to solve this type of problem by making certain assumptions which simplify the mathematical expressions and result in adequate accuracy unless the concentration of the weak acid or base is very small and the dissociation constant is very large. These assumptions are not necessary if one uses a Newton-Raphson iteration technique which allows for the solution of the third order equation without simplifications.

This calculator program allows the user to calculate the hydrogen ion concentration, $[H^+]$, and pH of a solution of a monoprotic weak acid if the ionization constant and concentration are known. For a solution of a weak base, the program will calculate the OH^- concentration and pOH once the ionization constant and concentration are entered. In addition, conversions from hydrogen ion concentration to pH or pOH and vice versa, and from pH to pOH and from $[H^+]$ to $[OH^-]$, are provided.

The following third order (cubic) equation is used

$$X^3 + K_a X^2 - (K_w + K_a C_a)X - K_w K_a = f(X) = 0 \qquad (K_b \text{ for bases})$$

where
$$X = [H^+] \text{ for acid, } [OH^-] \text{ for base}$$
$$K_a = \text{ionization constant of acid}$$
$$K_b = \text{ionization constant of base}$$
$$K_w = \text{ionization constant of water} = 10^{-14} \text{ at } 25°C$$
$$C_a \text{ or } C_b = \text{concentration (moles/liter) of acid or base}$$

The program uses the Newton-Raphson method of approximating the solution of a polynomial, where one evaluates $f(X)$ successively with approximate values of X. The first approximation of X is

$X = \sqrt{K_a C_a + K_w}$. Successive approximations are

$$X_{i+1} = X_i - \frac{f(X_i)}{f'(X_i)}$$

where $f'(X_i)$ is the first derivative of $f(X_i)$ evaluated at $X = X_i$:

$$f'(X) = 3X^2 + 2K_a X - (K_w + K_a C_a)$$

The calculation is iterated until X_{i+1} differs from X_i by 1% or less.

USER INSTRUCTIONS--RPN

Step	Instructions	Input	Keys	Output
1	Load program			
2	Optional: Enter molecular weight	M.W.	f c	M.W.
3	Calculate pH of weak acid solution:			
	Enter K_a directly	K_a	B	
	or after conversion of pK_a	pK_a	A B	
	or after conversion of K_b	K_b	f e	
			B	
	or after conversion of pK_b	pK_b	A	
			R/S	
			B	K_a
	Enter molar concentration directly	c_a	D	
	or after conversion from mg/ml	mg/ml	C D	pH*
4	Calculate pH of weak base solution:			
	Enter K_b directly	K_b	B	
	or after conversion of pK_b	pK_b	A B	
	or after conversion of K_a	K_a	f e	
			B	K_b
	Enter molar concentration directly	c_b	D	
	or after conversion from mg/ml	mg/ml	C D	pOH*
	Convert pOH to pH		R/S	pH
5	Convert pH to $[H^+]$	pH	A	$[H^+]$
	or pH to $[OH^-]$	pH	A	
			R/S	$[OH^-]$
	or pH to pOH	pH	f a	pOH
	To determine error of calculation		f b	$f(X)/f'(X)$

*$[H^+]$ or $[OH^-]$ is displayed for 1 sec, followed by pH or pOH

Line	Instructions	Input	Keys	Output
6	To recover pH or pOH		f d	pX
7	Convert pOH to pH or pK_b to pK_a (or reverse) or [H^+] to [OH^-] or K_a to K_b (or reverse)	pK_a [$\overline{OH}$]	f a f e	pK_b [H^+]

Program Listing--RPN

Line	Key	Comments	Line	Key	Comments
001	*LBLA		036	FIX	
002	CHS	Convert:	037	RTN	Display M.W.
003	10ˣ	pK_a to K_a	038	*LBLB	
004	SCI	pK_b to K_b	039	STO1	Enter K
005	R/S	pH to [H^+]	040	RTN	
006	*LBLe		041	*LBLc	Enter M.W.
007	EEX	Interchange:	042	STO6	
008	CHS	K_a and K_b or [H^+]	043	RTN	
009	1	and [OH^-]	044	*LBLD	Concentration → pH
010	4		045	RCL1	or pOH
011	X⇄Y		046	x	
012	÷		047	EEX	
013	SCI		048	CHS	
014	RTN		049	1	
015	*LBLE	Convert:	050	4	
016	LOG	[H^+] to pH	051	+	Calculate constants
017	CHS	[OH^-] to pOH	052	STO3	
018	FIX	K_a to pK_a	053	LSTX	
019	R/S		054	RCL1	
020	*LBLa		055	x	
021	1		056	STO4	
022	4	Interchange:	057	RCL3	
023	X⇄Y	pK_a and pK_b	058	√X	
024	-	pH and pOH	059	STO2	
025	FIX		060	GTO1	
026	RTN		061	RTN	
027	*LBLC		062	*LBLb	Display error
028	EEX	Calculate molarity	063	RCL5	→ f(X)/f'(X)
029	3	from mg/liter	064	SCI	
030	÷		065	R/S	Display K
031	RCL6		066	RCL1	
032	÷		067	RTN	
033	SCI		068	*LBL1	
034	R/S		069	RCL2	Newton-Raphson
035	RCL6		070	RCL1	iteration

Line	Key	Comments
071	+	
072	RCL2	
073	x	
074	RCL3	
075	-	
076	RCL2	
077	x	
078	RCL4	
079	-	
080	RCL2	
081	3	
082	x	
083	RCL1	
084	2	
085	x	
086	+	
087	RCL2	
088	x	
089	RCL3	
090	-	
091	÷	
092	STO5	
093	ABS	
094	RCL2	
095	9	
096	9	
097	÷	
098	X↕Y	Test approximations
099	X≤Y?	
100	GTOd	
101	RCL2	
102	RCL5	
103	-	
104	STO2	
105	GTO1	Calculate next
106	R/S	approximation
107	*LBLd	
108	RCL2	→ pH or pOH
109	SCI	Flash [H^+] or [OH^-]
110	PSE	
111	GTOE	

Register Contents, Labels, and Data Cards--RPN

Register	Contents	Labels	Contents
R_1	K	A	$pX \rightarrow X$
R_2	$[H^+]_{est}$	B	$K\uparrow$
R_3	$CK + K_w$	C	$mg/ml\uparrow$
R_4	KK_w	D	$C \rightarrow pH$
R_5	$f(X)/f'(X)$	E	$X \rightarrow$
R_6	M.W.	a	$pK_a \leftrightarrow pK_b$
		b	$\rightarrow$ error
		c	M.W.$\uparrow$
		d	$\rightarrow$ pH
		e	$X_a \leftrightarrow X_b$

User Instructions--Algebraic System

Step	Instructions	Input	Keys	Output
1	Enter K_a	K_a	2nd A'	
2	Optional: entry of pK_a	pK_a	2nd D'	K_a
3	Enter molar concentration	C_a	A	$[H^+]$
4	Optional: Enter molecular weight	M.W.	2nd C'	
	Enter mg to convert to concentration		R/S	C_a
	Enter concentration from display	C_a	A	$[H^+]$

Step	Instructions	Input	Keys	Output
5	Calculate pH	[H$^+$]	B	pH
6	Convert [H$^+$] to pH or pH to [OH$^-$]	pH	C	[H$^+$] or [OH$^-$]
7	Determine error of calculation		E	$f(X)/f'(X)$
8	Enter pOH: calculate pH or convert pK_b to pK_a or [H$^+$] to [OH$^-$]	pOH	2nd B'	pH

Program Listing--Algebraic System

Line	Key	Entry	Comments	Line	Key	Entry	Comments
000	76	LBL		026	00	00	
001	16	A'	Enter K_a	027	65	×	
002	42	STO		028	43	RCL	
003	00	00		029	04	04	
004	65	×		030	54	)	
005	01	1		031	85	+	
006	52	EE		032	01	1	
007	94	+/-		033	52	EE	
008	01	1		034	94	+/-	
009	04	4		035	01	1	
010	95	=		036	04	4	
011	42	STO	$K_a K_w$	037	95	=	X_0
012	02	02		038	34	ΓX	
013	91	R/S		039	42	STO	
014	76	LBL		040	05	05	
015	11	A	Enter concen-	041	53	(	
016	42	STO	tration	042	53	(	
017	04	04		043	43	RCL	
018	65	×		044	05	05	
019	43	RCL		045	85	+	
020	00	00		046	43	RCL	
021	95	=		047	00	00	
022	42	STO		048	54	)	
023	03	03	Newton-Raphson	049	65	×	
024	53	(	iteration	050	43	RCL	
025	43	RCL		051	05	05	

Line	Key	Entry	Comments	Line	Key	Entry	Comments
052	54	)		095	53	(	
053	75	-		096	43	RCL	
054	53	(		097	03	03	
055	43	RCL		098	75	-	
056	03	03		099	01	1	
057	75	-		100	52	EE	
058	01	1		101	94	+/-	
059	52	EE		102	01	1	
060	94	+/-		103	04	4	
061	01	1		104	54	)	
062	04	4		105	95	=	
063	54	)		106	42	STO	
064	54	)		107	07	07	
065	65	×		108	43	RCL	
066	43	RCL		109	05	05	$f'(X)$
067	05	05		110	55	÷	
068	54	)		111	09	9	
069	75	-		112	09	9	
070	43	RCL		113	95	=	
071	02	02		114	42	STO	
072	95	=		115	08	08	
073	42	STO	$f(X)$	116	43	RCL	
074	06	06		117	06	06	
075	53	(		118	55	÷	
076	53	(		119	43	RCL	
077	03	3		120	07	07	
078	65	×		121	95	=	
079	43	RCL		122	42	STO	
080	05	05		123	09	09	
081	54	)		124	43	RCL	$f(X)/f'(X)$
082	85	+		125	08	08	
083	53	(		126	32	X:T	
084	02	2		127	43	RCL	
085	65	×		128	09	09	
086	43	RCL		129	77	GE	
087	00	00		130	10	E'	Enter $[H^+]$
088	54	)		131	43	RCL	
089	54	)		132	05	05	
090	65	×		133	91	R/S	
091	43	RCL		134	76	LBL	
092	05	05		135	12	B	Calculate pH
093	54	)		136	28	LOG	
094	75	-		137	94	+/-	

Line	Key	Entry	Comments	Line	Key	Entry	Comments
138	42	STO		181	95	=	
139	10	10		182	91	R/S	
140	91	R/S		183	76	LBL	Enter pK_a
141	76	LBL		184	19	D'	Derive K_a
142	10	E'		185	94	+/-	
143	43	RCL	$X = X - \dfrac{f(X)}{f'(X)}$	186	22	INV	
144	05	05		187	28	LOG	
145	75	-		188	42	STO	
146	43	RCL		189	00	00	
147	09	09		190	91	R/S	Enter pH
148	95	=		191	76	LBL	
149	42	STO		192	13	C	Derive $[H^+]$
150	05	05		193	42	STO	
151	61	GTO		194	14	14	
152	00	00		195	94	+/-	
153	39	39		196	22	INV	
154	76	LBL	pOH → pH	197	28	LOG	
155	17	B'		198	91	R/S	
156	01	1		199	76	LBL	
157	04	4		200	15	E	Determine
158	75	-		201	43	RCL	$f(X)/f'(X)$
159	43	RCL		202	09	09	
160	10	10		203	91	R/S	
161	95	=					
162	91	R/S					
163	76	LBL	Enter M.W.				
164	18	C'	Enter mg/ml				
165	42	STO					
166	11	11					
167	91	R/S	Molar				
168	42	STO	concentration				
169	12	12					
170	65	×					
171	01	1					
172	52	EE					
173	94	+/-					
174	03	3					
175	95	=					
176	42	STO					
177	13	13					
178	55	÷					
179	43	RCL					
180	11	11					

Register Contents, Labels, and Data Cards--Algebraic System

Register	Contents	Labels	Contents
R0	K_a	Label A'	K_a
R02	$K_a K_w$	Label B'	pOH $\rightarrow$ pH
R03	CK_a	Label C'	M.W.
R04	Concentration	Label D'	$pK_a \rightarrow K_a$
R05	X_0	Label E'	$(Xn + 1)$
R06	$f(X)$	Label A	$C_a \rightarrow [H^+]$
R07	$f'(X)$	Label B	$[H^+] \rightarrow$ pH
R08	$X_0/99$	Label C	pH $\rightarrow [H^+]$
R09	$f(X)/f'(X)$	Label E	$f(X)/f'(X)$
R10	pH		
R11	M.W.		
R12	mg/ml		
R13	Molarity		
R14	$[H^+]$		

Example

A. Calculate the pH of 1.0×10^{-4} M solution of acetic acid.
The K_a of acetic acid is 1.8×10^{-5}.

Solution: Enter 1.8 EEX 5 CHS, press B $\to$ 1.8×10^{-5}

Enter 4,   press A $\to$ 1.0×10^{-4}

press D $\to$ 3.45×10^{-5}, 4.46

pH = 4.46

B. Calculate pH of a 3.0×10^{-6} M solution of NH_4Cl. The pK_a of ammonia is 4.75.

Solution: Enter 4.75,   press A $\to$ 1.78×10^{-5}

press R/S $\to$ 5.62×10^{-10}

press B $\to$ 5.62×10^{-10}

Enter 3 EEX 6 CHS,   press D $\to$ 1.08×10^{-7}, 6.97

pH = 6.97

References

1. Alan J. Rubin, "pH of Weak Acid/Base Solutions by Newton-Raphson Iteration," Hewlett-Packard *HP-67/97 User's Library Solutions in Chemistry*.

2. J.N. Butler (1964), *Ionic Equilibrium--A Mathematical Approach*, Addison-Wesley, Reading, Mass.

II

MACROMOLECULES IN SOLUTIONS

2A. PARTIAL SPECIFIC VOLUME BY AMINO ACID COMPOSITION

Apparent specific volumes of proteins can be estimated quite reliably from amino acid composition, since it has been found that the volumes of the macromolecules are closely equivalent to the sum of the individual amino acid residues:

$$\bar{\upsilon} = \frac{\sum_i V_i W_i}{\sum_i W_i}$$

where W_i = percent by weight of ith residue
         V_i = specific volume of ith residue
             = $1/\rho$ of ith residue
         $V_i W_i$ = percent volume of ith residue
         ρ = density

One needs to know the total number of amino acid residues, as well as the molecular weight of the protein.

Although this method for determining the apparent specific volume for proteins is not a rigorous one, it has been found for many proteins that the values calculated in this manner are in close agreement with the experimentally determined ones. Table 1 lists the amino acids, their densities, and their partial specific volumes. Also listed are the molecular weights. These values are necessary for the calculation of the partial specific volume of a protein from the amino acid composition.

TABLE 1

Amino Acid	Molecular Weight	Density (gm/ml)	V_i (ml/g)
Alanine	89.1	1.401	0.714
Arginine	174.2	1.1	0.909
Aspartic acid	133.1	1.66	0.602
Asparagine	132.2	1.66	0.602
Glutamic acid	147.1	1.538	0.650
Glutamine	146.1	1.49	0.67
Glycine	75.1	1.601	0.625
Leucine	131.2	1.165	0.858
Isoleucine	131.2	1.165	0.858
Tyrosine	181.2	1.456	0.687
Valine	117.1	1.23	0.813
Methionine	149.2	1.34	0.746
Serine	105.1	1.537	0.651
Cysteine	121.2	1.587	0.63
Histidine	155.2	1.493	0.67
Phenylalanine	165.2	1.299	0.77
Proline	115.1	1.316	0.76
Threonine	119.1	1.464	0.683
Lysine	146.2	1.22	0.82

User Instructions--RPN

Step	Instructions	Input	Keys	Output
1	Initialize: clear registers		f c	0.00
2	Enter molecular weight of protein calculated from amino acid composition	$M.W._{prot}$	f a	$M.W._{prot}$
3	Enter: number of residues of ith amino acid M.W. of ith amino acid specific volume of ith amino acid	n_i $M.W._i$ V_i	ENTER↑ ENTER↑ A	 i
4	Calculate and print partial specific volume of protein		B	$\bar{\upsilon}$
5	Optional: Output of: $\sum W_i$ $\sum V_i W_i$ $\sum n_i$		C D E	$\sum W$ $\sum V_i W_i$ $\sum n_i$

Program Listing--RPN

Line	Key	Comments
01♦LBL e		
02 7CLREG		Initialize:
03 CLX		Clear registers
04 7DSP2		
05 RTN		
06♦LBL a		
07 STO 20		Store $M.W._{prot}$
08 RTN		
09♦LBL A		
10 STO 04		$\rightarrow$ store $\bar{V}_i$
11 RDN		
12 18		$\rightarrow$ $(M.W._i$ - 1 mole
13 -		$H_2O)$
14 X<>Y		
15 ST+ 00		$\rightarrow \sum n_i$
16 *		
17 RCL 20		
18 /		
19 E2		
20 *		
21 ST+ 01		$\rightarrow \sum w_i$
22 RCL 04		
23 *		
24 ST+ 02		$\rightarrow \sum v_i w_i$
25 1		
26 ST+ 03		$\rightarrow \sum i$
27 RCL 03		
28 RTN		

Line	Key	Comments
29♦LBL B		
30 RCL 02		Calculate $\bar{v}$
31 RCL 01		
32 /		
33 7DSP3		
34 7PRTX		$\rightarrow \bar{v}$
35 ADV		
36 RTN		
37♦LBL C		
38 RCL 01		
39 7PRTX		$\rightarrow \sum w_i$
40 RTN		
41♦LBL D		
42 RCL 02		
43 7PRTX		$\rightarrow \sum v_i w_i$
44 ADV		
45 RTN		
46♦LBL E		
47 RCL 00		
48 7PRTX		$\rightarrow \sum n_i$
49 RTN		
50 STOP		

Register Contents, Labels, and Data Cards--RPN

Register	Contents	Labels	Contents
R_0	$\sum n_i$	A	$n_i \uparrow M.W._i \uparrow V_i$
R_1	$\sum W_i$	B	$\rightarrow \bar{\upsilon}$
R_2	$\sum V_i W_i$	C	$\rightarrow \sum W_i$
R_3	$\sum n_i$	D	$\rightarrow \sum V_i W_i$
R_A	$M.W._{prot.}$	E	$\rightarrow \sum n_i$
		a	$M.W._{prot.}$
		e	Initialize

User Instructions--Algebraic System

Step	Instructions	Input	Keys	Ouput
1	Print option		2nd B'	
2	Enter molecular weight of protein	$M.W._{prot}$	2nd A'	
3	Enter number of residues of ith amino acid	n_i	A	
4	Enter molecular weight of ith amino acid	$M.W._i$	B	
5	Enter specific volume of ith amino acid	$\bar{V}_i$	C	
6	Calculate partial specific volume		D	$\bar{\upsilon}$

Program Listing--Algebraic System

Line	Key	Entry	Comments	Line	Key	Entry	Comments
000	76	LBL		041	12	12	
001	17	B'	Print toggle	042	42	STO	Print M.W.$_i$
002	87	IFF		043	29	29	
003	08	08		044	02	2	
004	45	Y$^\times$		045	02	2	
005	86	STF		046	71	SBR	
006	08	08		047	85	+	
007	92	RTN		048	53	(	
008	76	LBL		049	43	RCL	
009	45	Y$^\times$		050	12	12	
010	22	INV		051	75	-	
011	86	STF		052	01	1	
012	08	08		053	08	8	
013	92	RTN		054	54	)	
014	76	LBL	M.W.$_{prot}$	055	65	$\times$	
015	16	A'		056	43	RCL	
016	42	STO		057	11	11	
017	10	10		058	54	)	
018	42	STO	Print M.W.$_{prot}$	059	55	÷	
019	29	29		060	43	RCL	
020	02	2		061	10	10	
021	00	0		062	54	)	
022	71	SBR	Number of	063	65	$\times$	
023	85	+	amino acid	064	01	1	Calculate Sums
024	91	R/S	residues	065	00	0	
025	76	LBL		066	00	0	
026	11	A	Print n	067	95	=	
027	42	STO		068	36	PGM	Enter V_i
028	11	11		069	01	01	
029	42	STO		070	32	X:T	
030	29	29		071	91	R/S	Print V_i
031	02	2		072	76	LBL	
032	01	1		073	13	C	
033	71	SBR	Enter M.W.$_i$	074	42	STO	
034	85	+		075	13	13	
035	43	RCL		076	42	STO	
036	29	29		077	29	29	
037	91	R/S		078	02	2	
038	76	LBL		079	03	3	
039	12	B		080	71	SBR	
040	42	STO		081	85	+	

Line	Key	Entry	Comments
082	43	RCL	
083	29	29	
084	78	Σ+	
085	91	R/S	
086	76	LBL	
087	14	D	Calculate $\bar{v}$
088	43	RCL	
089	06	06	
090	55	÷	
091	43	RCL	
092	04	04	
093	95	=	
094	42	STO	
095	29	29	
096	02	2	
097	04	4	Print $\bar{v}$
098	71	SBR	
099	85	+	
100	91	R/S	
101	76	LBL	
102	85	+	
103	87	IFF	
104	08	08	
105	95	=	
106	43	RCL	
107	29	29	
108	92	RTN	
109	76	LBL	
110	95	=	
111	42	STO	
112	25	25	
113	73	RC*	
114	25	25	
115	69	OP	
116	04	04	
117	43	RCL	
118	29	29	
119	69	OP	
120	06	06	
121	69	OP	
122	00	00	
123	98	ADV	
124	92	RTN	

Register Contents, Labels, and Data Cards--Algebraic System

Register	Contents	Labels	Contents
R01	$\sum y$	Label A'	Enter M.W.$_{prot}$
		Label B'	Print routine
R02	$\sum y$	Label A	Enter n_i
R03	n	Label B	Enter M.W.$_i$
R04	$\sum x$	Label C	Enter V_i
R05	$\sum x^2$	Label D	Calculate $\overline{v}$
R06	$\sum xy$		
R10	M.W.$_{prot}$		
R11	N_i		
R12	M.W.$_i$		

Alpha print code	Register	Contents
30433300.	20	MWP
31351736.	21	NRES
30431313.	22	MWAA
42360000.	23	VS
42330000.	24	VP

Example

The following data concerning the enzyme ribonuclease are from a paper by C.W. Hirs, W.H. Stein and S. Moore, in *Journal of Biological Chemistry 211*, 941 (1954). The molecular weight of the protein calculated from these data is 13,886. These data are necessary to calculate the partial specific volume of ribonuclease from its amino acid composition.

Amino Acid	Number of Residues per Molecule	W_i	V_i	$V_i W_i$
Aspartic acid	7	5.8	0.620	3.49
Asparagine	9	7.40	0.602	4.45
Glutamic acid	4	3.72	0.65	2.42
Glutamine	8	7.38	0.67	4.94
Glycine	3	1.23	0.625	0.77
Alanine	12	6.14	0.714	4.39
Valine	9	6.42	0.813	5.22
Leucine	2	1.63	0.858	1.40
Isoleucine	3	2.45	0.858	2.10
Serine	15	9.41	0.651	6.13
Threonine	10	7.28	0.683	4.97
Cysteine	8	5.95	0.63	3.75
Methionine	4	3.78	0.746	2.82
Proline	5	3.50	0.76	2.66
Phenylalanine	3	3.18	0.77	2.45
Tyrosine	6	7.05	0.687	4.84
Histidine	4	3.95	0.67	2.65
Lysine	10	9.23	0.82	7.57
Arginine	4	4.5	0.909	4.09
	$\sum n_i = 126$	$\sum W_i = 100$		$\sum V_i W_i = 71.11$

Solution

$$\bar{v} = \frac{7.11}{100} = 0.711$$

References

1. H.K. Schachman, "Ultracentrifugation, Diffusion, and Viscosity, in *Methods in Enzymology IV*, 32-103.

2. E.J. Cohn, and J.T. Edsall Eds. (1943), *Proteins, Amino Acids, and Peptides*, Reinhold Publishing Corporation, New York.

3. C.W. Hirs, W.H. Stein, and S. Moore (1954), *Journal of Biological Chemistry 211*, 941.

2B. MOLECULAR WEIGHT FROM OSMOTIC PRESSURE DATA

One of the simpler ways to estimate the molecular weight of a molecule is by osmotic pressure measurements. The presence of solute (in most biochemical applications, a protein) in a solution has the effect of lowering the chemical potential (activity) of the solvent (usually water). The difference between the chemical potential of a pure solvent and that of a solvent-solute mixture is measured by a device called an osmometer. Typically, an osmometer consists of two compartments, one containing solvent and one containing solute plus solvent, separated by a partition (a differentially permeable membrane) which is permeable only to solvent. Since the membrane allows solvent flow but restricts solute passage, solvent will move to the compartment containing solute. The osmotic pressure can be thought of as the pressure required to equalize the chemical potentials of the two compartment system (i.e., the pressure which will prevent solvent movement).

In a dilute solution the osmotic pressure π is equal to the gas constant R times the absolute temperature T and n, the number of moles of solute divided by the volume of the solution:

$$\pi = \frac{RTn}{V} \tag{1}$$

Remembering that the number of moles is equal to the number of grams per molecular weight, equation 1 can be rearranged in the form of a straight line where C = concentration in grams per liter:

$$\frac{\pi}{CRT} = \text{slope } C + \frac{1}{\text{M.W.}}$$

Since this equation applies to dilute solutions, it becomes necessary to measure the osmotic pressure for a number of concentrations of the protein in question. The slope of the line in equation 2 then represents solvent-solute interactions (departure from dilute solution ideality), while the reciprocal of the slope intercept gives the molecular weight.

Approximating the molecular weight by osmotic measurements is accurate provided that the sample is homogeneous. However, if the sample is heterogeneous, the above procedure will yield the average molecular weights of the components in the sample.

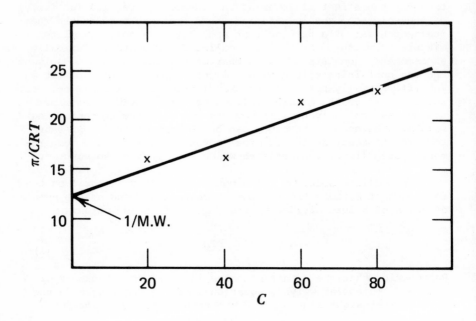

Figure 2B.   Osmotic pressure data plot

User Instructions--RPN

Step	Instructions	Input	Keys	Output
1	Initialize: Clear registers		f e	0.000
2	To set print flag		f d	1.00
3	To clear print flag		f d	0.00
4	Enter Temperature (°C)	T	f a	T
5	Enter: protein concentration	C	ENTER↑	C
	osmotic pressure (atm)	π	A	n
	To delete incorrect data:			
6	Enter: protein concentration	C	ENTER↑	C
	osmotic pressure	π	B	$n - 1$
	Repeat step 5 for all data			
7	Calculate coefficient of determination and molecular weight		C	r^2 M.W.
	To compute best fit straight line for plotting:			
8	Enter: protein concentration (gm/liter)	C	D	π/CRT

Program Listing--RPN

Line	Key	Comments	Line	Key	Comments
001	*LBLe		013	+	
002	CLRG	Initialize: Clear	014	.	
003	P≷S	registers	015	0	
004	CLRG		016	8	
005	CLX		017	2	
006	FIX		018	1	
007	DSP3		019	x	Compute RT (°K)
008	RTN		020	STO0	
009	*LBLα	Enter T (°C)	021	CLX	
010	2		022	RTN	
011	7		023	*LBLA	
012	3		024	RCL0	Enter C

Line	Key	Comments	Line	Key	Comments
025	÷		072	+	
026	X⇄Y	Enter π	073	÷	
027	÷	→ π/CRT	074	PRTX	→ r^2
028	LSTX		075	RCL6	
029	F2?	Print if print	076	RCL4	
030	GSB9	flag is on	077	RCLB	
031	Σ+	Accumulate sums	078	X	
032	RTN		079	–	
033	*LBLB		080	RCL9	
034	RCL0		081	÷	
035	÷		082	STOA	→ intercept
036	X⇄Y		083	1/X	
037	÷		084	DSP0	
038	LSTX	Delete incorrect	085	PRTX	→ Molecular weight
039	F2?	data	086	SPC	
040	GSB3		087	P⇄S	
041	F2?		088	RTN	
042	GSB9		089	*LBLD	Plotting/projection
043	Σ–		090	ENG	routine
044	RTN		091	DSP3	
045	*LBLC		092	RCLB	
046	P⇄S		093	X	
047	RCL8		094	RCLA	
048	RCL4		095	+	
049	RCL6		096	RTN	
050	X	Compute regression	097	*LBLd	Print flag: Set and
051	RCL9	coefficients	098	0	Clear
052	÷		099	F2?	
053	–		100	RTN	
054	ENT↑		101	1	
055	ENT↑		102	SF2	
056	RCL4		103	RTN	
057	X²		104	*LBL9	Print subroutine
058	RCL9		105	PRTX	
059	÷		106	X⇄Y	
060	RCL5		107	ENG	
061	X⇄Y		108	PRTX	Print π/CRT in
062	–		109	SF2	engineering nota
063	÷		110	SPC	tion
064	STOB	→ slope	111	X⇄Y	
065	X		112	FIX	
066	RCL6		113	RTN	
067	X²		114	*LBL3	Print deletion
068	RCL9		115	SPC	indicator
069	÷		116	DSP0	
070	CHS		117	1	
071	RCL7		118	CHS	

Line	Key	Comments
119	PRTX	
120	SF2	
121	.DSP3	
122	R↓	
123	RTN	
124	R/S	

Register Contents, Labels, and Data Cards--RPN

Register	Contents	Labels	Contents
R_0	RT	A	$C \uparrow \pi (+)$
R_{s4}	$\sum C$	B	$C \uparrow \pi (-)$
R_{s5}	$\sum C^2$	C	$\rightarrow r^2$, M.W.
R_{s6}	$\sum \pi/CRT$	D	$C \rightarrow \pi/CRT$
R_{s7}	$\sum (\pi/CRT)^2$	a	T (°C)
R_{s8}	$\sum C(\pi/CRT)$	d	Print?
R_{s9}	n	e	Initialize
R_A	1/M.W.		
R_B	Slope		

User Instructions--Algebraic System

Step	Instructions	Input	Keys	Output
1	Print option		2nd B'	
2	Enter Temperature (°C)	T(°C)	2nd A'	
3	Enter protein concentration (gm/liter)	C	A	
4	Enter osmotic pressure (atm)	π	B	π/CRT
5	Enter data into linear regression		R/S	n
6	Calculate molecular weight		C	M.W.
7	Calculate slope		R/S	slope
8	Calculate correlation coefficient		D	r
9	Enter C'; derive π/CRT'	C'	E	π/CRT

Program Listing--Algebraic System

Line	Key	Entry	Comments	Line	Key	Entry	Comments
000	76	LBL		041	85	+	
001	17	B'		042	43	RCL	
002	87	IFF	Print toggle	043	12	12	
003	08	08		044	36	PGM	
004	45	Y×		045	01	01	Linear
005	86	STF		046	32	X¦T	regression
006	08	08		047	92	RTN	
007	92	RTN		048	76	LBL	Enter π
008	76	LBL		049	12	B	
009	45	Y×		050	42	STO	
010	22	INV		051	13	13	
011	86	STF		052	42	STO	
012	08	08	*Program note	053	29	29	
013	92	RTN		054	01	1	Call print
014	76	LBL	$T(°C)$	055	08	8	routine
015	16	A'		056	71	SBR	
016	42	STO		057	85	+	
017	10	10		058	43	RCL	
018	42	STO		059	13	13	
019	29	29	Call print	060	55	÷	
020	01	1	routine	061	53	(	
021	06	6		062	93	.	
022	71	SBR		063	00	0	
023	85	+		064	08	8	→ R
024	85	+		065	02	2	
025	02	2		066	01	1	
026	07	7		067	65	×	
027	03	3		068	43	RCL	
028	95	=		069	11	11	
029	42	STO		070	65	×	
030	11	11	Enter C	071	43	RCL	Calculate
031	91	R/S		072	12	12	π/CRT
032	76	LBL		073	95	=	
033	11	A		074	57	ENG	
034	42	STO		075	42	STO	
035	12	12		076	13	13	
036	42	STO		077	42	STO	Call print
037	29	29	Call print	078	29	29	routine
038	01	1	routine	079	01	1	
039	07	7		080	09	9	
040	71	SBR		081	71	SBR	

*Note: RTN is keyed in as INV SBR

Line	Key	Entry	Comments	Line	Key	Entry	Comments
082	85	+		123	71	SBR	
083	91	R/S		124	85	+	
084	43	RCL		125	91	R/S	$C' \rightarrow \pi/CRT$
085	13	13		126	76	LBL	
086	22	INV		127	15	E	
087	57	ENG		128	42	STO	
088	78	Σ+		129	15	15	
089	92	RTN		130	69	OP	
090	76	LBL		131	14	14	
091	13	C		132	57	ENG	
092	69	OP		133	91	R/S	Print sub-
093	12	12		134	76	LBL	routine
094	95	=		135	85	+	
095	35	1/X	M.W.	136	87	IFF	
096	42	STO		137	08	08	
097	29	29	Call print	138	95	=	
098	02	2	routine	139	43	RCL	
099	01	1		140	29	29	
100	71	SBR		141	92	RTN	
101	85	+		142	76	LBL	
102	91	R/S		143	95	=	
103	69	OP		144	42	STO	
104	12	12		145	25	25	
105	32	X:T	Slope	146	73	RC*	
106	42	STO		147	25	25	
107	29	29		148	69	OP	
108	02	2	Call print	149	04	04	
109	02	2	routine	150	43	RCL	
110	71	SBR		151	29	29	
111	85	+		152	69	OP	
112	91	R/S		153	06	06	
113	76	LBL		154	69	OP	
114	14	D		155	00	00	
115	69	OP	r	156	98	ADV	
116	13	13		157	92	RTN	
117	22	INV		158	76	LBL	
118	57	ENG		159	75	-	
119	42	STO		160	87	IFF	
120	29	29	Call print	161	08	08	
121	02	2	routine	162	55	÷	
122	03	3		163	43	RCL	
				164	29	29	

Line	Key	Entry	Comments		Line	Key	Entry	Comments
165	92	RTN			178	43	RCL	
166	76	LBL			179	29	29	
167	55	÷			180	99	PRT	
168	42	STO			181	98	ADV	
169	26	26			182	92	RTN	
170	73	RC*						
171	26	26						
172	69	OP						
173	04	04						
174	69	OP						
175	05	05						
176	69	OP						
177	00	00						

Register Contents, Labels, and Data Cards--Algebraic System

Register	Contents	Labels	Contents
R0 → R06	Linear regression	Label A'	Enter $T(°C)$
		Label B'	Print toggle
R10	$T(°C)$	Label A	Enter C
R11	$T_{abs}(°K)$	Label B	Enter π
R12	C	Label C	Calculate π/CRT
R13	π	Label D	r
R15	C'	Label E	$C' \to \pi/CRT$

Data Card (4 2nd write)

Alpha Print Code	Register	Contents
37173033.	16	TEMP
15323115.	17	CONC
76323630.	18	OSM
76631535357.	19	π/CRT
30430000.	21	MW
36273317.	22	SLPE
15323535.	23	CORR

Example

An electrophoretically homogeneous protein was dissolved in 0.1 M buffer-0.2 M salt solution of neutral pH (the isoelectric pH of the protein), and measurements of osmotic pressure as a function of protein concentration were made at 5°C. Calculate the molecular weight of the protein.

Table I

Protein Concentration (gm/liter)	Osmotic Pressure (atm)
20	0.0072
40	0.0147
60	0.0298
80	0.0421

Solution

C (gm/liter)	π/CRT
20	15.77×10^{-6}
40	16.10×10^{-6}
60	21.76×10^{-6}
80	23.06×10^{-6}

M.W.   = 81,332
Slope = 137.56×10^{-9}
r = 0.941

Fitted line

c'	π/CRT'
20	15.05×10^{-6}
80	23.30×10^{-6}

References

1. C. Tanford (1967), *Physical Chemistry of Macromolecules*, John Wiley & Sons, Inc. New York, pp. 217-221.

2C. INTRINSIC VISCOSITY FROM VISCOMETRIC DATA

The viscosity of a solution is an index of the internal re-
sistance of the fluid to flow (a measure of energy dissipation in
flow). Viscosity measurements can provide estimates of parameters which lead to information on macromolecular shape. Determination of the viscosity of a particular solution is approached
by comparing the solution viscosity η with the viscosity of the
pure solvent η_0, and is expressed in equation 1 as the relative
viscosity η_r:

$$\eta_r = \frac{\eta}{\eta_0} \tag{1}$$

Changes in the viscosity of a solution produced by adding
solute (i.e., the volume fraction times an unknown numerical factor V plus the higher terms $K\phi^2 + \ldots$, which are solute-solute
interactions) are given by the specific viscosity η_{sp} (equation
2), which is equal to the relative viscosity minus 1:

$$\eta_{sp} = \eta_r - 1 = \frac{\eta_s - \eta_0}{\eta_0} = V_\phi + K_\phi^2 + \ldots \tag{2}$$

Since it is more convenient to work with concentration terms,
equation 2 can be rearranged to yield equation 3, where the volume fraction is replaced by the concentration of solute C, in
grams per milliliter, and the specific volume $\bar{\upsilon}$:

$$\frac{\eta_{sp}}{C} = \bar{\upsilon}V + KV^2 C \ldots \tag{3}$$

By graphical solution of equation 3 (Figure 2C) one can estimate
the intrinsic viscosity $[\eta]$ from the slope intercept (in effect,
taking the limit η_{sp}/C as $C \to 0$) and thereby eliminate interaction effects. After determination of the intrinsic viscosity and
with estimates of the protein specific volume $\bar{\upsilon}$, the amount of
water affiliated with the protein (i.e., water of hydration) δ
and the solvent density $\rho_{20,w}$, the viscosity number ν (numerical
factor) can be determined as follows:

$$[\eta] = \nu V = \nu(\bar{\upsilon} + \frac{\delta}{\rho_{20,w}}) \tag{4}$$

$$\nu = \frac{[\eta]}{\overline{\upsilon} + \delta/\rho_{20,w}} \tag{5}$$

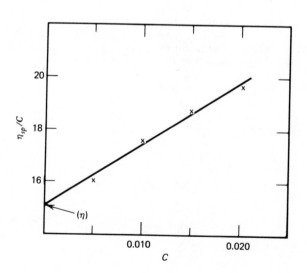

Figure 2C. A plot of η_{sp}/C versus C to determine $[\eta]$

User Instructions--RPN

Step	Instructions	Input	Keys	Output
1	Initialize: Clear Registers		f e	0.00
2	To set print flag		f d	1.00
3	To clear print flag		f d	0.00
4	Enter partial specific volume (ml/gm)	$\overline{\upsilon}$	ENTER↑	
5	Enter water of hydration (ml/mg)	δ	f a	
6	Enter flow time for solvent (sec)	$F.T._s$	ENTER↑	$F.T._s$
7	Enter density of solvent	ρ_s	f b	ρ_s
8	Enter: protein concentration (mg/ml)	C	ENTER↑	C
	flow time (sec)	F.T.	A	n
	To delete incorrect data			
9	Enter: protein concentration	C	ENTER↑	C
	flow time	F.T.	B	$n-1$
	Repeat step 8 for all data			
10	Compute coefficient of determination and intrinsic viscosity		C	r^2 $[\eta]$
11	To compute viscosity factor		D	ν
	To compute best fit straight line:			
12	Enter protein concentration	C	E	η_{sp}/C

Program Listing--RPN

Line	Key	Comments
001	*LBLe	
002	CLRG	Initialize: Clear
003	P≵S	Registers
004	CLRG	
005	CLX	
006	RTN	
007	*LBLa	
008	.	
009	9	
010	9	
011	8	$\rightarrow \delta/\rho_{20,w}$
012	÷	
013	X≵Y	
014	+	$\rightarrow \bar{\upsilon} + \delta/\rho_{20,w}$
015	LSTX	
016	ST00	Store $\bar{\upsilon}$
017	R↓	Store $\bar{\upsilon} + \delta/\rho_{20,w}$
018	ST01	
019	RTN	
020	*LBLb	
021	1/X	$\rightarrow 1/\rho_{solvent}$
022	RCL0	
023	-	
024	ST03	$\rightarrow 1/\rho_{solvent} - \bar{\upsilon}$
025	R↓	
026	ST02	$\rightarrow$ F.T.$_{solvent}$
027	RTN	
028	*LBLA	
029	RCL2	
030	÷	
031	X≵Y	
032	ENT↑	
033	ENT↑	
034	RCL3	
035	x	
036	1	
037	+	
038	R↑	
039	x	
040	1	
041	-	
042	X≵Y	
043	÷	$\rightarrow C$
044	LSTX	$\rightarrow \eta_{sp}/C$
045	F2?	

Line	Key	Comments
046	GSB9	Print if print flag
047	Σ+	is set
048	RTN	Accumulate sums
049	*LBLB	
050	RCL2	Delete incorrect
051	÷	data
052	X≵Y	
053	ENT↑	
054	ENT↑	
055	RCL3	
056	x	
057	1	
058	+	
059	R↑	
060	x	
061	1	
062	-	
063	X≵Y	
064	÷	
065	LSTX	
066	F2?	
067	GSB3	
068	F2?	
069	GSB9	
070	Σ-	
071	RTN	
072	*LBLC	
073	P≵S	
074	RCL8	
075	RCL4	
076	RCL6	
077	x	Compute intrinsic
078	RCL9	viscosity
079	÷	
080	-	
081	ENT↑	
082	ENT↑	
083	RCL4	
084	X²	
085	RCL9	
086	÷	
087	RCL5	
088	X≵Y	
089	-	
090	÷	

Line	Key	Comments
091	STOB	
092	x	
093	RCL6	
094	x^2	
095	RCL9	
096	÷	
097	CHS	
098	RCL7	
099	+	
100	÷	
101	PRTX	→ r^2
102	RCL6	
103	RCL4	
104	RCLB	
105	x	
106	-	
107	RCL9	
108	÷	
109	STOA	
110	PRTX	→ [η]
111	SPC	
112	P⇄S	
113	RTN	
114	*LBLD	Compute viscosity
115	RCLA	factor
116	RCL1	
117	÷	
118	PRTX	→ ν
119	SPC	
120	RTN	
121	*LBLE	Plotting/projection
122	RCLB	
123	x	
124	RCLA	
125	+	
126	RTN	
127	*LBL9	Print subroutine
128	PRTX	
129	X⇄Y	
130	PRTX	
131	X⇄Y	
132	SF2	
133	SPC	
134	RTN	
135	*LBL3	Print deletion
136	SPC	indicator
137	DSP0	

Line	Key	Comments
138	1	
139	CHS	
140	PRTX	
141	SF2	
142	DSP3	
143	R↓	
144	RTN	
145	*LBLd	Print flag: Set
146	0	and Clear
147	F2?	
148	RTN	
149	1	
150	SF2	
151	RTN	

Register Contents, Labels, and Data Cards--RPN

Register	Contents	Labels	Contents
R_0	$\bar{v}$	A	$C \uparrow \text{F.T.}(+)$
R_1	$\bar{v} + \delta/\rho_{20,w}$	B	$C \uparrow \text{F.T.}(-)$
R_2	F.T._s	C	$\rightarrow r^2,\ [\eta]$
R_3	$1/\rho_s - \bar{v}$	D	$\rightarrow \nu$
R_{s4}	$\sum C$	E	$C' \rightarrow \eta_{sp}/C'$
R_{s5}	$\sum C^2$	a	$\bar{v} \uparrow \delta$
R_{s6}	$\sum \eta_{sp}/C$	b	$\text{F.T.}_s \uparrow \rho_s$
R_{s7}	$\sum (\eta_{sp}/C)^2$	d	Print?
R_{s8}	$\sum C(\eta_{sp}/C)$	e	Initialize
R_{s9}	n		
R_A	$[\eta]$		
R_B	Slope		

User Instructions--Algebraic System

Step	Instruction	Input	Keys	Output
1	Enter partial specific volume of protein (ml/gm)	$\bar{v}$	2nd A'	
2	Enter hydration volume (ml/gm protein)	δ	2nd B'	$(\bar{v} + \delta/\rho_{20,w})$
3	Enter outflow time of solvent	F.T._s	2nd C' .	
4	Enter density of solvent	ρ_s	2nd D'	$(1/\rho_s - \bar{v})$

Step	Instruction	Input	Keys	Output
5	Enter concentration of protein solution (mg/ml)	C	A	
6	Enter outflow time (sec) for protein solution to obtain η_{sp}/C (gm/ml)	F.T.	B	η_{sp}/C
7	Enter linear regression subroutine		R/S	
	Repeat steps 5, 6, and 7 for all data pairs			
8	Calculate intrinsic viscosity		C	$[\eta]$
9	Calculate correlation coefficient		R/S	r
10	Derive η_{sp}/C from C' for fitted line	C'	D	η_{sp}/C'
11	Calculate viscosity number		R/S	ν

Program Listing—Algebraic System

Line	Key	Entry	Comments	Line	Key	Entry	Comments
000	76	LBL		015	53	(	
001	16	A'		016	43	RCL	
002	42	STO	$\bar{\nu}$	017	10	10	
003	10	10		018	95	=	
004	91	R/S		019	42	STO	$\to \bar{\nu} + \delta/\rho_{20,w}$
005	76	LBL	Enter δ	020	12	12	
006	17	B'		021	91	R/S	
007	42	STO		022	76	LBL	Enter flow
008	11	11		023	18	C'	time for sol-
009	55	÷		024	42	STO	vent
010	93	.		025	13	13	
011	09	9		026	91	R/S	Enter density
012	09	9		027	76	LBL	of solvent ρ_s
013	08	8		028	19	D'	
014	85	+	$\to \delta/\rho_{20,w}$	029	42	STO	

Line	Key	Entry	Comments
030	14	14	
031	35	1/X	
032	75	-	$(1/\rho_s - \bar{\upsilon})$
033	43	RCL	
034	10	10	
035	95	=	
036	42	STO	
037	15	15	
038	91	R/S	
039	76	LBL	Enter concen-
040	11	A	tration C
041	42	STO	
042	16	16	
043	76	LBL	Linear
044	32	X:T	regression
045	36	PGM	
046	01	01	
047	32	X:T	
048	92	RTN	Enter flow
049	76	LBL	time F.T.
050	12	B	
051	42	STO	
052	17	17	
053	55	÷	F.T./F.T.$_s$
054	43	RCL	
055	13	13	
056	65	×	
057	53	(	
058	01	1	
059	85	+	
060	53	(	
061	43	RCL	ρ_m/ρ_s
062	15	15	
063	65	×	
064	43	RCL	
065	16	16	
066	95	=	
067	42	STO	ρ_m/ρ_s
068	18	18	
069	53	(	
070	43	RCL	
071	18	18	

Line	Key	Entry	Comments
072	75	-	
073	01	1	$(\eta_r - 1)$
074	54	)	
075	55	÷	
076	43	RCL	η_{sp}/C
077	16	16	
078	95	=	
079	91	R/S	
080	78	Σ+	
081	92	RTN	
082	76	LBL	
083	13	C	
084	69	OP	$[\eta]$
085	12	12	
086	42	STO	
087	19	19	
088	91	R/S	
089	69	OP	
090	13	13	r
091	91	R/S	
092	76	LBL	
093	14	D	
094	42	STO	$C' \rightarrow \eta_{sp}/C$
095	20	20	
096	69	OP	
097	14	14	
098	91	R/S	
099	43	RCL	
100	19	19	
101	55	÷	υ
102	43	RCL	
103	12	12	
104	95	=	
105	91	R/S	

Register Contents, Labels, and Data Cards--Algebraic System

Register	Contents	Labels	Contents
R0 → R06	Linear regression	Label A	Enter C
R10	$\overline{\upsilon}$	Label B	Enter F.T.
R11	δ	Label C	Calculate $[\eta]$
R12	$\overline{\upsilon} + \delta/\rho_{20,w}$	Label D	Enter F.T.
R13	$F.T._s$	Label A'	Enter $\overline{\upsilon}$
R14	ρ_s	Label B'	Enter δ
R15	$(1/\rho_s - \overline{\upsilon})$	Label C'	Enter F.T.
R16	C	Label D'	Enter ρ_s
R17	F.T.		
R18	ρ_m/ρ_s		
R19	η_{sp}/C		
R20	$[\eta]$		

Example

Using the measurements of protein concentration and the corresponding outflow times from a capillary viscometer, determine the intrinsic viscosity and the viscosity number for the protein.

Concentration of protein (gm/ml)	Outflow Time (sec)
0.020	298
0.015	274
0.010	252
0.005	232

$\bar{\upsilon}$ = 0.725 ml/gm (partial volume of specific protein)
δ = 0.4 cm^3/gm of protein (water of hydration)
$F.T._s$ = 215 sec
$\rho_{20,s}$ = 1.002 gm/ml
$\rho_{20,w}$ = 0.998 gm/ml

Solution

C	η_{sp}/C
0.005	16.11
0.010	17.53
0.015	18.64
0.020	19.68

$\lim c \to 0$: η_{sp}/C = intercept = $[\eta]$ = 15.03
r = 0.997

Fitted line

c'	η_{sp}/c'
0.005	16.22
0.020	19.76

Viscosity number = ν = 13.35 cm^3/gm

References

1. K.E. Van Holde (1971), *Physical Biochemistry*, Prentice-Hall, Inc., Englewood Cliffs, N.J.

2D. MOLECULAR WEIGHT FROM GEL FILTRATION

Gel filtration is a technique which allows the separation of differing molecules on the basis of size. Gels consist of polymers of carbohydrate (*Sephadex*) or acrylamide *(Biogel)* which are packed in a column to function as a molecular sieve. The sample is applied to the column in a small volume of solvent and is eluted with excess solvent. In this type of experimental procedure, the porous gels will exclude larger molecules, resulting in their elution from the column before smaller molecules in the sample solution. Since molecular size is related to molecular weight, one may determine the molecular weight of a molecule by measuring its elution volume relative to the values for other marker proteins (proteins of known molecular weight). The elution of proteins in aqueous medium can be equated to the molecular weight as follows:

$$\frac{V_e}{V_0} = K \log \text{M.W.} \tag{1}$$

Here V_e/V_0 is the ratio of the relative elution volume V_e for the particular protein in question to the column void volume V_0 (the dead volume or the solvent volume between the gel particles as determined by elution of a high molecular weight polysaccharide—*blue dextran*—which is totally excluded from the gel). The relative elution volume is equal to a column constant K times the log of the molecular weight. The calculator program allows the entry of V_0 followed by entry of the elution volumes for the marker proteins which span the molecular weight range of interest. The molecular weight of the unknown protein can then be determined by substitution of the observed elution volume in the equation for the straight line derived from the linear regression analysis of the marker proteins.

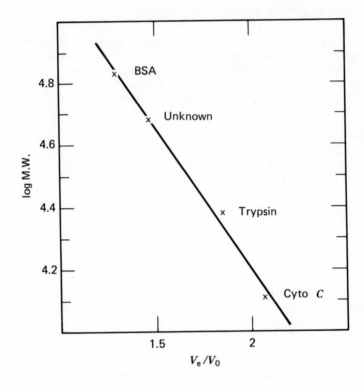

Figure 2D. Plot of log M.W. versus the relative elution volume

User Instructions--RPN

Step	Instructions	Input	Keys	Output
1	Initialize: Clear registers		f e	0.00
2	To set print flag		f d	1.00
3	To clear print flag		f d	0.00
4	Enter void volume of column	V_0	f a	V_0
5	Enter elution volume of reference protein	V_e	ENTER↑	V_e
6	Enter molecular weight of reference protein	M.W.	A	n
	To delete incorrect data:			
7	Enter: V_e molecular weight	V_e M.W.	ENTER↑ B	$\cdot V_e$ $n - 1$
8	Repeat steps 5 and 6 for all reference proteins			
9	Compute coefficient of determination		C	r^2
10	Enter elution volume of protein of unknown molecular weight	V_e'	D	M.W.

Program Listing--RPN

Line	Key	Comments	Line	Key	Comments
001	*LBLe		046	RCL5	
002	CLRG	Initialize: clear	047	X≠Y	
003	P≠S	registers	048	-	
004	CLRG		049	÷	
005	CLX		050	STOB	
006	DSP2		051	x	
007	RTN		052	RCL6	
008	*LBLa		053	X²	
009	STO0	Store V_0	054	RCL9	
010	RTN		055	÷	
011	*LBLA		056	CHS	
012	LOG	→ log M.W.	057	RCL7	
013	X≠Y		058	+	
014	RCL0	→ V_e/V_0	059	÷	
015	÷		060	STOC	
016	F2?	Print if print flag	061	RCL6	
017	GSB9	is on	062	RCL4	
018	Σ+	Accumulate sums	063	RCLB	
019	RTN		064	x	
020	*LBLB		065	-	
021	LOG	Delete incorrect	066	RCL9	
022	X≠Y	data	067	÷	
023	RCL0		068	STOA	
024	÷		069	P≠S	
025	F2?		070	RCLC	
026	GSB3		071	PRTX	→ r^2
027	F2?		072	RTN	
028	GSB9		073	*LBLD	
029	Σ-		074	RCL0	
030	RTN		075	÷	
031	*LBLC		076	RCLB	Compute molecular
032	P≠S		077	x	weight of unknown
033	RCL8	Compute regression	078	RCLA	
034	RCL4	coefficients	079	+	
035	RCL6		080	DSP0	
036	x		081	10ˣ	
037	RCL9		082	PRTX	
038	÷		083	RTN	
039	-		084	*LBLd	
040	ENT↑		085	0	
041	ENT↑		086	F2?	
042	RCL4		087	RTN	
043	X²		088	1	Pring flag: Set and
044	RCL9		089	SF2	Clear
045	÷		090	RTN	

Line	Key	Comments
091	*LBL9	Print subroutine
092	PRTX	
093	X⇄Y	
094	PRTX	
095	X⇄Y	
096	SF2	
097	SPC	
098	RTN	
099	*LBL3	Print deletion
100	SPC	indicator
101	DSP0	
102	1	
103	CHS	
104	PRTX	
105	SF2	
106	DSP2	
107	R↓	
108	RTN	

Register Contents, Labels, and Data Cards--RPN

Register	Contents	Labels	Contents
R_0	V_0	A	$V_e \uparrow$ M.W.(+)
R_{s4}	$\sum V_e/V_0$	B	$V_e \uparrow$ M.W.(-)
R_{s5}	$\sum (V_e/V_0)^2$	C	$\rightarrow r^2$
R_{s6}	$\sum \log$ M.W.	D	$V_e' \rightarrow$ M.W.'
R_{s7}	$\sum (\log \text{M.W.})^2$	a	$V_0 \uparrow$
R_{s8}	$\sum (V_e/V_0) \times (\log \text{M.W.})$	d	Print?

Register	Contents	Labels	Contents
R_{s9}	n	e	Initialize
R_A	Intercept		
R_B	Slope		
R_C	r^2		

User Instructions--Algebraic System

Step	Instructions	Input	Keys	Output
1	Print toggle		2nd B'	
2	Enter void volume	V_0	2nd A'	
3	Enter elution volume of known protein	V_e	A	V_e/V_0
4	Enter V_e/V_0 into linear regression		R/S	
5	Enter molecular weight of known protein	M.W.	B	log M.W.
6	Enter log M.W. into linear regression		R/S	n
7	Calculate correlation coefficient		C	r
8	Enter elution volume of unknown protein	V_e	D	V_e/V_0
9	Derive log molecular weight of unknown protein		R/S	log M.W.
10	Derive molecular weight of unknown protein		E	M.W.

Program Listing--Algebraic System

Line	Entry	Key	Comments	Line	Key	Entry	Comments
000	76	LBL		041	42	STD	
001	17	B'	Print toggle	042	29	29	
002	87	IFF		043	01	1	Call print
003	08	08		044	06	6	subroutine
004	45	Y×		045	71	SBR	
005	86	STF		046	75	-	
006	08	08		047	43	RCL	
007	·92	RTN		048	29	29	
008	76	LBL		049	91	R/S	
009	45	Y×		050	36	PGM	Linear regres-
010	22	INV		051	01	01	sion
011	86	STF		052	32	X!T	
012	08	08		053	91	R/S	
013	92	RTN		054	76	LBL	
014	76	LBL		055	12	B	Enter M.W.
015	16	A'		056	42	STD	
016	42	STD	Enter V_0	057	29	29	
017	10	10		058	01	1	
018	42	STD		059	07	7	
019	29	29		060	71	SBR	Call print
020	01	1		061	85	+	subroutine
021	04	4		062	28	LDG	
022	71	SBR	Call print	063	42	STD	
023	85	+	subroutine	064	12	12	
024	91	R/S		065	42	STD	
025	76	LBL		066	29	29	
026	11	A		067	01	1	
027	42	STD		068	08	8	
028	11	11	Enter V_e	069	71	SBR	Call print
029	42	STD		070	75	-	subroutine
030	29	29		071	43	RCL	
031	01	1		072	12	12	
032	05	5		073	91	R/S	
033	71	SBR		074	78	Σ+	
034	85	+	Call print	075	42	STD	
035	43	RCL	subroutine	076	29	29	
036	29	29		077	01	1	Linear regres-
037	55	÷		078	09	9	sion
038	43	RCL		079	71	SBR	
039	10	10		080	85	+	Call print
040	95	=	V_e/V_0	081	91	R/S	subroutine

Line	Key	Entry	Comments	Line	Key	Entry	Comments
082	76	LBL		123	08	8	
083	13	C		124	71	SBR	
084	69	OP		125	75	-	
085	13	13	r	126	43	RCL	
086	42	STO		127	29	29	
087	29	29		128	91	R/S	
088	02	2		129	76	LBL	
089	00	0	Call print	130	15	E	
090	71	SBR	subroutine	131	22	INV	M.W.
091	85	+		132	28	LOG	
092	43	RCL		133	42	STO	
093	29	29		134	29	29	
094	91	R/S		135	02	2	Call print
095	76	LBL	Enter V'_e	136	02	2	subroutine
096	14	D		137	71	SBR	
097	42	STO		138	85	+	
098	29	29	Call print	139	91	R/S	
099	02	2	subroutine	140	76	LBL	
100	01	1		141	85	+	Print subrou-
101	71	SBR		142	87	IFF	tine
102	85	+		143	08	08	
103	43	RCL		144	95	=	
104	29	29		145	43	RCL	
105	55	÷		146	29	29	
106	43	RCL	V_e/V'_0	147	92	RTN	
107	10	10		148	76	LBL	
108	95	=		149	95	=	
109	42	STO	Call print	150	42	STO	
110	29	29	subroutine	151	23	23	
111	01	1		152	73	RC*	
112	06	6		153	23	23	
113	71	SBR		154	69	OP	
114	75	-		155	04	04	
115	43	RCL		156	43	RCL	
116	29	29		157	29	29	
117	91	R/S		158	69	OP	
118	69	OP		159	06	06	
119	14	14	Derive log	160	69	OP	
120	42	STO	M.W.	161	00	00	
121	29	29		162	98	ADV	
122	01	1	Call print	163	92	RTN	
			subroutine				

Line	Key	Entry	Comments
164	76	LBL	
165	75	-	
166	87	IFF	
167	08	08	
168	55	÷	
169	43	RCL	
170	29	29	
171	92	RTN	
172	76	LBL	
173	55	÷	
174	42	STD	
175	24	24	
176	73	RC*	
177	24	24	
178	69	OP	
179	04	04	
180	69	OP	
181	05	05	
182	69	OP	
183	00	00	
184	43	RCL	
185	29	29	
186	99	PRT	
187	98	ADV	
188	92	RTN	

Register Contents, Labels, and Data Cards--Algebraic System

Register	Contents	Labels	Contents
R0 → R06	Linear regression	Label A'	Enter V_0
		Label B'	Print toggle
R10	V_0	Label A	Enter V_e
R11	V_e	Label B	Enter M.W.
R12	log M.W.	Label C	Derive r
R13	M.W. unknown	Label D	Enter V_e'; derive V_e/V_0'
		Label E	Derive M.W. unknown

Data Card (4 wnd Write)

Alpha Print Code	Register	Contents
42320000.	14	V□
42540000.	15	Ve
4254634232.	16	Ve/V□
30430000.	17	MW
2732223043.	18	LOGMW
31000000.	19	N
15323535.	20	CORR
42546500.	21	Ve"
30430041.	22	MW U

Example

The elution volume for an unknown protein from a gel filtration column was 50 ml. Using the elution volumes in Table 1 and the void volume (V_0 = 35 ml), determine the molecular weight of the unknown protein.

Table I

Protein	V_e (ml)	Molecular Weight
Cytochrome c	72.3	13,000
Trypsin	64.8	24,000
BSA	45.5	67,000
Unknown protein	51.5	--

Solution

Protein	V_e/V_0	Log M.W.
Cytochrome c	2.07	4.11
Trypsin	1.85	4.38
BSA	1.3	4.83
Unknown	1.47	4.68

$$r = 0.99$$
$$\text{M.W.}_{unknown} = 48,140.9$$

References

1. H. Determann (1968), *Gel Chromatography,* Springer Verlag, New York, pp. 105-111.

2E.   DIFFUSION COEFFICIENT

The diffusion of macromolecules can be thought of as the rate of movement of mass down a concentration gradient. Fick's first law (equation 1) can be used to describe the phenomenon. The flux--(mass per unit time) is equal to the concentration per transfer distance times the area A and the proportionality constant (diffusion coefficient, D):

$$J = -DA \frac{\partial c}{\partial x} \tag{1}$$

One way to determine the diffusion coefficient is by substitution of the continuity equation (equation 2) into Fick's first law:

$$\frac{\partial c}{\partial t} = - \frac{\partial J}{\partial x} \tag{2}$$

to give Fick's second law:

$$\frac{\partial c}{\partial t} = D \frac{\partial^2 c}{\partial x^2} \tag{3}$$

Equation 3 can be solved by specifying the initial and boundary conditions where diffusion occurs in an infinite system (i.e., free diffusion where the diffusion coefficient is independent of concentration). The solution to equation 3 is then expressed as the derivative, which is a probability equation:

$$\frac{dc}{dx} = \frac{c_0}{2(\pi Dt)^{\frac{1}{2}}} e^{-x^2/4Dt} \tag{4}$$

Thus concentration gradients can be monitored as a function of time by analysis of the family of Gaussian curves (Figure 2E1) described by equation (4).

The height of the curve is given by the maximum at $x = 0$, while c_0 is the initial concentration difference across the boundary (equation 5), and the area under the curve is equivalent to the initial concentration (equation 6):

$$\left(\frac{dc}{dx} \right)_{x=0} = \frac{c_0}{2(\pi Dt)^{\frac{1}{2}}} = H \tag{5}$$

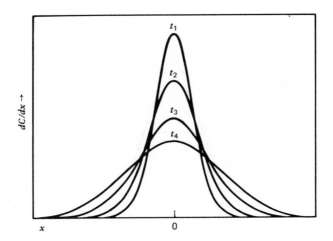

Figure 2E1. Example of a family of Gaussian Curves

$$C_0 = \int_{-\infty}^{\infty} \frac{\partial C}{\partial x}\, dx = A \tag{6}$$

By combining equations 5 and 6, a graphical solution (Figure 2E2) for the determination of the diffusion coefficient based on constant area, curve height decline, and time can be obtained.

Although the height-area method is accurate for the determination of the diffusion coefficient, it requires many experimental measurements (i.e., one must determine the area under the curve by Simpson's rule, as well as measuring the height change as a function of time). A more rapid alternative method which requires fewer measurements is the width method (1). The simple width method is based on the observation that there is a fixed relationship between the area and the width of a Gaussian curve (2). The submaximal height h_i is obtained from equation 2 by using any X_i value other than zero:

$$\left(\frac{dc}{dx}\right) X_i = He^{-x_i^2/4Dt} = h_i \tag{7}$$

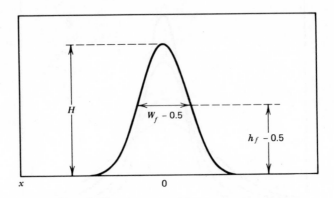

Figure 2E2.   Measurement of diffusion data using width method

The full width W_i can be substituted into equation 7
($W_i = 2x_i$) to obtain the relation between width, h_i, D, and time.
By designating f_i as the relative fractional height ($f = h_i/H$)
expressions 8 and 9 can be formulated to solve for the diffusion
coefficient:

$$W^2 = -16Dt \ \ln f \tag{8}$$

$$D = \frac{- \text{ slope}}{16 \ \ln f} \tag{9}$$

The calculator program incorporates the ultracentrifuge schlieren
pattern emperical correction factor F_c, the optical system factor
F_m, and a conversion factor (6×10^3) which alters equation 9 to
equation 10:

$$D = - \frac{\text{slope } f_c^2}{16 F_m^2 \ (\ln f) \ 6 \times 10^3} \tag{10}$$

The diffusion coefficient of the molecule at the temperature of
the experiment is then corrected to standard conditions using
equation 11:

$$D_{20,w} = D_{obs}\left(\frac{293}{273 + T}\right)\left(\frac{\eta_{T,w}}{\eta_{20,w}}\right)\left(\frac{\eta_{solv}}{\eta_w}\right) \qquad (11)$$

where D_{obs} = measured diffusion coefficient

T = temperature of the diffusion experiment (°C)

(η_{solv}/η_w) = the relative viscosity of the solvent to that of water

$\eta_{T,w}$ = viscosity of water at the temperature of the experiment

$\eta_{20,w}$ = viscosity of water at 20°C

User Instructions--RPN

Step	Instructions	Input	Keys	Output
1	Initialize: Clear Registers		f e	0.00
2	To set print flag		f d	1.00
3	To clear print flag		f d	0.00
4	Enter correction factor	F_c	ENTER↑	F_c
5	Enter magnification factor	F_m	ENTER↑	F_m
6	Enter fractional value of height	f	f a	
7	Enter experimental temperature (°C)	T	ENTER↑	T
8	Enter viscosity of water at temperature of experiment	η_T	ENTER↑	η_T
9	Enter relative viscosity	η_{solv}/η_w	f c	η_{solv}/η_w
10	Enter: time (min) width (mm) at h_i	t W	ENTER↑ A	t n
	To delete incorrect data:			
11	Enter: time width	t W	ENTER↑ B	t $n-1$
12	Compute coefficient of determination and diffusion coefficient		C	r^2 D
13	To correct diffusion coefficient relative to water at 20°C		D	$D_{20,w}$
	To compute best fit straight line:			
14	Enter time, t	t	E	W^2

Program Listing--RPN

Line	Key	Comments	Line	Key	Comments
001	*LBLa		045	F2?	
002	LN	Enter:	046	GSB3	Print deletion indi-
003	STO2	correction factor	047	F2?	cator if print flag
004	R↓	magnification factor	048	GSB9	is set
005	X²	fractional height	049	X⇄Y	Delete data
006	STO1		050	Σ-	
007	R↓		051	RTN	
008	X²		052	*LBLC	Compute regression
009	STO0		053	SPC	coefficients
010	RTN	Enter temperature of	054	FIX	
011	*LBLb	experiment and vis-	055	P⇄S	
012	1	cosity of water at	056	RCL8	
013	.	temperature of exper-	057	RCL4	
014	0	ment	058	RCL6	
015	0		059	x	
016	2		060	RCL9	
017	÷		061	÷	
018	STO4		062	-	
019	R↓		063	ENT↑	
020	2		064	ENT↑	
021	7		065	RCL4	
022	3		066	X²	
023	+		067	RCL9	
024	2		068	÷	
025	9		069	RCL5	
026	3		070	X⇄Y	
027	÷		071	-	
028	1/X		072	÷	
029	STO5		073	STOB	
030	RTN		074	x	
031	*LBLc		075	RCL6	
032	STO3	Enter relative	076	X²	
033	RTN	viscosity	077	RCL9	
034	*LBLA		078	÷	
035	X²	Enter time (min)	079	CHS	Coefficient of deter-
036	CF3	and width (mm)	080	RCL7	mination
037	F2?		081	+	
038	GSB9	Print data if print	082	÷	
039	X⇄Y	flag is set	083	STOC	→ r^2
040	Σ+	Accumulate sums	084	PRTX	
041	RTN		085	RCL6	
042	*LBLB	Deletion of data	086	RCL4	
043	X²		087	RCLB	
044	SF3		088	x	

Line	Key	Comments	Line	Key	Comments
089	–		135	SPC	
090	RCL9		136	RTN	
091	÷		137	*LBL3	
092	STOA		138	SPC	Deletion flag
093	RCLB		139	DSP0	indicator
094	CHS		140	1	
095	P⇄S		141	CHS	
096	RCL0		142	PRTX	
097	x		143	SF2	
098	RCL1		144	DSP2	
099	RCL2		145	R↓	
100	9		146	RTN	
101	6		147	*LBLe	
102	EEX		148	FIX	Initialize: clear
103	3	Uncorrected diffu-	149	DSP2	registers
104	x	sion coefficient	150	CLRG	
105	x		151	P⇄S	
106	÷		152	CLRG	
107	DSP3	→ D	153	CLX	
108	SCI		154	RTN	
109	PRTX		155	*LBLE	
110	SPC		156	FIX	Plotting/projection
111	RTN		157	DSP2	using best fit param-
112	*LBLD	Corrected diffusion	158	RCLB	eters
113	RCL3	coefficient	159	X⇄Y	
114	RCL4		160	x	
115	RCL5		161	LSTX	
116	x		162	X⇄Y	
117	x		163	RCLA	
118	x	→ $D_{20,w}$	164	+	
119	PRTX		165	F2?	
120	SPC		166	GSB9	
121	RTN		167	RTN	
122	*LBLd	Print flag: Set and			
123	0	Clear			
124	F2?				
125	RTN				
126	1				
127	SF2				
128	RTN	Print subroutine			
129	*LBL9				
130	X⇄Y				
131	PRTX				
132	X⇄Y				
133	PRTX				
134	SF2				

Register Contents, Labels, and Data Cards--RPN

Register	Contents		Labels	Contents
R_0	F_c^2		A	$t \uparrow W(+)$
R_1	F_m^2		B	$t \uparrow W(-)$
R_2	$\ln f$		C	$\rightarrow r^2, D$
R_3	η_{solv}/η_w		D	$\rightarrow D_{20,w}$
R_4	$\eta_{T,w}/\eta_{20,w}$		E	$t \rightarrow W^2$
R_5	$293/T$ (°K)		a	$F_c \uparrow F_m \uparrow f$
R_{s4}	$\sum t$		b	T (°C) $\uparrow \eta_T$
R_{s5}	$\sum t^2$		c	$\eta_{solv}/\eta_w \uparrow$
R_{s6}	$\sum W^2$		d	Print?
R_{s7}	$\sum W^4$		e	Initialize
R_{s8}	$\sum W^2 t$			
R_{s9}	n			
R_A	Intercept			
R_B	Slope			
R_C	r^2			

User Instructions--Algebraic System

Step	Instructions	Input	Keys	Output
1	Enter schlieren diffusion pattern correction factor	F_c	2nd A'	F_c^2
2	Enter magnification factor for optical system	F_m	2nd B'	F_m^2
3	Enter relative fractional height		2nd C'	ln
4	Enter temperature (°C)	$T(°C)$	2nd D'	$293/T$
5	Enter viscosity of water at temperature of diffusion experiment	$\eta_{T,w}$	2nd E'	$\eta_{T,w}/\eta_{20,w}$
6	Enter viscosity of water at 20°C	$\eta_{20,w}$	R/S	
7	Enter viscosity of solvent at 20°C	η_{solv}	R/S	η_{solv}/η_w
8	Enter time	t	A	
9	Enter width of curve	W	B	W^2
10	Enter linear regression Enter all t, before D calculation		R/S	n
11	Calculate diffusion coefficient		C	D
12	Calculate correction coefficient		R/S	r
13	Derive W^2 from t'	t'	D	W^2
14	Calculate diffusion coefficient for standard conditions		E	$D_{20,w}$

Program Listing--Algebraic System

Line	Key	Entry	Comments	Line	Key	Entry	Comments
000	76	LBL		041	55	÷	
001	16	A'	F_c	042	01	1	
002	33	X²		043	93	.	
003	42	STO	F_c^2	044	00	0	
004	10	10		045	00	0	
005	91	R/S		046	02	2	
006	76	LBL	F_m	047	95	=	
007	17	B'		048	42	STO	
008	33	X²		049	23	23	$n_{T,w}/n_{20,w}$
009	42	STO	F_m^2	050	91	R/S	
010	11	11		051	42	STO	
011	91	R/S	f	052	24	24	$n_{20,w}$
012	76	LBL		053	91	R/S	
013	18	C'		054	42	STO	
014	23	LNX	$\ln f$	055	25	25	n_{solv}
015	42	STO		056	55	÷	
016	12	12		057	43	RCL	
017	91	R/S		058	24	24	
018	76	LBL	$T(°C)$	059	95	=	
019	19	D'		060	42	STO	$n_{solv}/n_{20,w}$
020	85	+		061	26	26	
021	02	2		062	91	R/S	
022	07	7		063	76	LBL	Enter t
023	03	3		064	11	A	
024	95	=		065	42	STO	
025	42	STO		066	13	13	
026	20	20		067	36	PGM	Call linear re-
027	02	2		068	01	01	gression
028	09	9		069	32	X:T	
029	03	3		070	92	RTN	
030	55	÷		071	76	LBL	
031	43	RCL		072	12	B	Enter W
032	20	20		073	33	X²	
033	95	=		074	42	STO	
034	42	STO	$293/T$	075	14	14	
035	21	21		076	91	R/S	$\to W^2$
036	91	R/S		077	78	Σ+	
037	76	LBL		078	92	RTN	
038	10	E'		079	76	LBL	Compute regres-
039	42	STO	$n_{T,w}$	080	13	C	sion coefficients
040	22	22		081	69	OP	

Line	Key	Entry	Comments		Line	Key	Entry	Comments
082	12	12			123	43	RCL	
083	32	X:T			124	21	21	
084	94	+/-			125	65	×	
085	65	×			126	43	RCL	
086	43	RCL			127	23	23	
087	10	10			128	65	×	
088	55	÷			129	43	RCL	
089	53	(			130	26	26	$D_{20,w}$
090	43	RCL			131	95	=	
091	11	11			132	91	R/S	
092	65	×						
093	53	(						
094	43	RCL						
095	12	12						
096	65	×						
097	09	9						
098	06	6						
099	00	0						
100	00	0						
101	00	0	D					
102	54	)						
103	95	=						
104	52	EE	r					
105	42	STO						
106	15	15						
107	91	R/S						
108	69	OP						
109	13	13						
111	76	LBL	$t' \to w^2$					
112	14	D						
113	42	STO						
114	16	16						
115	69	OP						
116	14	14						
117	91	R/S						
118	76	LBL						
119	15	E						
120	43	RCL						
121	15	15						
122	65	×						

Register Contents, Labels, and Data Cards--Algebraic System

Register	Contents	Labels	Contents
R0 → R6	Linear regression sums	Label A'	F_c
R10	F_c^2	Label B'	F_m
R11	F_m^2	Label C'	f
R12	$\ln f$	Label D'	$T(^\circ C)$
R15	r	Label E'	n_T
R21	$293/T(^\circ K)$	Label A	t
R23	$n_{T,w}/n_{20,w}$	Label B	W
R24	$n_{20,w}$	Label C	D
R26	$n_{solv}/n_{20,w}$	Label D	$t' \rightarrow W^2$
		Label E	$D_{20,w}$

Example

The data presented here for analysis are extracted from Figure 4 of reference 3.   The correction factors were also taken from reference 3.

Native α-glycerophosphate dehydrogenase (α-GDH) from rabbit muscle was centrifuged in a double-sector synthetic boundary cell in a Model E ultracentrifuge at a speed (5200 rpm) low enough that no significant sedimentation occurred.   The width values of the schlieren patterns recorded photographically are reported below at 8 min intervals.   Calculate the diffusion coefficient from the data at the three fractional heights.

$$F_m \quad = 2.155$$
$$T \quad = 5°C$$
$$\eta_T \quad = 1.519 \text{ centipoises, the viscosity of water at } 5°C$$
$$\eta_{solv}/\eta_w = 1.013, \text{ the relative viscosity}$$

$f = 0.3$ $F_C = 1.020$		$f = 0.5$ $F_C = 1.025$		$f = 0.7$ $F_C = 1.018$	
t (min)	W (mm)	t (min)	W (mm)	t (min)	W (mm)
8	1.8	8	1.5	8	1.0
24	2.6	16	1.7	24	1.4
32	2.8	24	1.9	32	1.6
40	3.2	32	2.2	40	1.7
48	3.4	40	2.4	48	1.8
56	3.6	48	2.6	56	1.9
64	3.8	56	2.7	64	2.1
72	4.0	64	2.8	72	2.2
80	4.2	72	3.1	80	2.3
88	4.4	80	3.2	88	2.4
		88	3.2		

Solution

r^2	$= 0.998$	r^2	$= 0.993$	r^2	$= 0.995$
D	$= 3.849 \times 10^{-7}$	D	$= 3.791 \times 10^{-7}$	D	$= 3.861 \times 10^{-7}$
$D_{20,w}$	$= 6.230 \times 10^{-7}$	$D_{20,w}$	$= 6.136 \times 10^{-7}$	$D_{20,w}$	$= 6.250 \times 10^{-7}$

Average $D_{20,w} = 6.21 \times 10^{-7}$ cm^2/sec

Area/height method (3) gives a value of 6.08×10^{-7} cm^2/sec

References

1. K.E. Van Holde (1971), *Physical Biochemistry*, Prentice-Hall, Inc., Englewood Cliffs, N.J., pp. 85-97.

2. Henry B. Bull (1964), *An Introduction to Physical Biochemistry*, F.A. Davis Company, Philadelphia, Pa., Chapter 11.

3. G.J. Wei and W.C. Deal, Jr. (1978), "A Simple Width Method for Easy Determination of Diffusion Coefficients with Maximum Utilization of Data," *Analytical Biochemistry 87*, 433-446.

III

SEDIMENTATION

3A. ROUTINE CENTRIFUGATION CALCULATIONS

Biochemists and molecular biologists make frequent use of the centrifuge, which is one of the most important instruments in the laboratory. The centrifugal force exerted by the centrifuge is given by

$$F = \omega^2 rm \tag{1}$$

where m is the mass at a distance r from the center of rotation. Since ω, the angular velocity, is to be expressed in radians per second and a radian is a ratio, not a length, it is clear that equation 1 has the proper dimensions (see Appendix II).

Since one revolution is equal to 2π radians, the number of revolutions per second (rps) is equal to 2π radians per second, and equation 1 becomes

$$F = 4\pi^2 r (\text{rps})^2 m \tag{2}$$

The relative centrifugal force (R.C.F.) expressed as multiples of the force of gravity is

$$\text{R.C.F.} = \frac{4\pi^2 r (\text{rps})^2}{gm} = \frac{4\pi^2 r (\text{rps})^2}{g} \tag{3}$$

$$= 0.0402 r (\text{rps})^2 \tag{4}$$

The number 0.0402 is the result of combining all constants in equation 3.   The value of g used in the computation is 980 cm/sec^2 at sea level.

The program presented here is really two programs in one. The first uses the interchangeable solutions algorithm and the data entry sensing flag of the HP-67.   The TI-59 does not have a data entry flag, so the program for that machine uses a different method.   The interchangeable solution algorithm allows the user to solve for either the revolutions per minute or the centrifugal force (in g's), given the other known quantity.

The second program allows the user to determine the revolutions per minute required to generate a sucrose density gradient equivalent to a standard sucrose density gradient at a given rpm and time.   This program is useful when one must, for one reason or another, conduct a sucrose density gradient experiment for a period of time greater or less than the time used in a previous experiment which resulted in a satisfactory sucrose density gradient centrifugation.

Table 1 provides minimum, maximum, and average radii of several commonly used centrifuge rotors manufactured by Beckman Instruments.   For other rotors manufactured by either Beckman or other companies, consult the appropriate catalogs and other literature.

TABLE 1. SOME COMMON BECKMAN CENTRIFUGE ROTORS

Rotor Type[*]		Radii (cm)		
		r_{min}	r_{max}	r_{av}
Fixed angle:	65	3.7	7.8	5.7
	50 Ti	3.8	8.1	5.9
	40	3.8	8.1	5.9
	30	5.0	10.5	7.8
Swinging bucket:	SW 50.1	5.97	10.73	8.35
	SW 27	7.5	16.1	11.8

*Rotor designations are Beckman's (e.g., 50 Ti refers to the type 50 titanium rotor.)

User Instructions--RPN

Step	Instructions	Input	Keys	Output
1	Load program			
2	Enter radius of rotor (cm)	r	A	r
3	Enter desired rpm	rpm	B	rpm
4	Compute R.C.F. (in g's)		C	R.C.F.
	or			
5	Enter radius of rotor (cm)	r	A	r
6	Enter desired R.C.F. (in g's)	R.C.F.	C	R.C.F.
7	Compute required rpm		B	rpm
	For sucrose density gradient centrifugation:			
8	Enter time of reference run (hr)	t_1	f a	t_1
9	Enter rpm of reference run	rpm_1	f b	rpm_1
10	Enter time of new run (hr)	t_2	f c	t_2
11	Compute rpm required to give a centrifugation run equivalent to the reference run		f d	rpm_2

Program Listing--RPN

Line	Key	Comments
001	*LBLA	
002	DSP2	Radius input?
003	STOA	If not, compute
004	F3?	radius and store in
005	R/S	R_A
006	*LBL1	
007	RCLB	
008	GSBe	
009	.	
010	0	
011	4	
012	0	
013	2	
014	x	
015	RCLC	
016	X≠Y	
017	÷	
018	STOA	
019	R/S	
020	*LBLB	RPM input?
021	STOB	If not, compute rpm
022	F3?	and store in R_B
023	R/S	
024	*LBL2	
025	RCLC	
026	.	
027	0	
028	4	
029	0	
030	2	
031	RCLA	
032	x	
033	÷	
034	√X	
035	6	
036	0	
037	x	
038	STOB	
039	DSP0	
040	R/S	
041	*LBLC	
042	STOC	
043	F3?	
044	R/S	

Line	Key	Comments
045	*LBL3	
046	.	
047	0	
048	4	Relative centrifugal
049	0	force input?
050	2	If not, compute
051	RCLA	R.C.F. and store in
052	x	R_C
053	RCLB	
054	GSBe	
055	x	
056	STOC	
057	DSP0	
058	R/S	
059	*LBLa	
060	DSP0	Time of reference
061	STOD	run input?
062	F3?	If not, compute t_1
063	R/S	and store in R_D
064	*LBL4	
065	RCL3	
066	GSBe	
067	RCL2	
068	x	
069	RCLE	
070	GSBe	
071	÷	
072	STOD	
073	R/S	
074	*LBLb	RPM of reference run
075	STOE	input?
076	F3?	If not, compute rpm
077	R/S	and store in R_E ;
078	*LBL5	
079	RCL3	
080	GSBe	
081	RCL2	
082	x	
083	RCLD	
084	÷	
085	√X	
086	6	
087	0	
088	x	

Line	Key	Comments
089	STOE	
090	DSP0	
091	R/S	
092	*LBLc	
093	STO2	Time of new run
094	F3?	input?
095	R/S	If not, compute time
096	*LBL6	of new run and store
097	RCLE	in R_2
098	GSBe	
099	RCLD	
100	x	
101	RCL3	
102	GSBe	
103	÷	
104	STO2	
105	DSP0	
106	R/S	
107	*LBLd	
108	STO3	RPM of new run
109	F3?	input?
110	R/S	If not, compute re-
111	*LBL7	quired rpm of new
112	RCLE	run and store in R_3
113	GSBe	
114	RCLD	
115	x	
116	RCL2	
117	÷	
118	√x	
119	6	
120	0	
121	x	
122	STO3	
123	DSP0	

Register Contents, Labels, and Data Cards--RPN

Register	Contents	Labels	Contents
R_2	$time_2$	A	$\leftrightarrow$ radius
R_3	rpm_2	B	$\leftrightarrow$ rpm
R_A	radius of rotor (cm)	C	$\leftrightarrow$ R.C.F.
R_B	rpm	a	$\leftrightarrow t_1$
R_C	R.C.F. (relative centrifugal force)	b	$\leftrightarrow rpm_1$
R_D	$time_1$	c	$\leftrightarrow t_2$
R_E	rpm_1	d	$\leftrightarrow rpm_2$

User Instructions--Algebraic System

Step	Instructions	Input	Keys	Output
1	Enter radius of rotor (cm)	r	A	
2	Enter rpm; calculate relative centrifugal force	rpm	B	R.C.F.
3	Enter radius of rotor (cm)	r	A	
4	Enter relative centrifugal force	R.C.F.	C	rpm
	For sucrose density gradient centrifugation:			
1	Enter radius of rotor (cm)	r	A	
2	Enter rpm of reference run	rpm	B	r.c.f.
3	Enter time of reference run (hr)	t_1	D	
4	Enter time of new run; calculate rpm to give equivalent run	t_2	E	rpm_2

Program Listing--Algebraic System

Line	Key	Entry	Comments	Line	Key	Entry	Comments
000	76	LBL		041	43	RCL	
001	11	A	r	042	00	00	
002	42	STO		043	95	=	
003	00	00		044	34	$\lceil$X	
004	91	R/S		045	65	×	
005	76	LBL		046	06	6	
006	12	B	rpm	047	00	0	rpm
007	42	STO		048	95	=	
008	01	01		049	42	STO	
009	55	÷		050	01	01	t_1
010	06	6		051	91	R/S	
011	00	0		052	76	LBL	
012	95	=		053	14	D	
013	33	X²	rps^2	054	42	STO	
014	42	STO		055	04	04	
015	02	02		056	65	×	
016	65	×		057	43	RCL	
017	43	RCL		058	03	03	
018	00	00		059	95	=	
019	65	×		060	42	STO	t_2
020	93	.		061	05	05	
021	00	0		062	91	R/S	
022	04	4		063	76	LBL	
023	00	0		064	15	E	
024	02	2		065	42	STO	
025	95	=	R.C.F.	066	06	06	
026	42	STO		067	43	RCL	
027	03	03		068	05	05	
028	91	R/S		069	55	÷	
029	76	LBL	R.C.F.	070	43	RCL	
030	13	C		071	06	06	
031	42	STO		072	95	=	
032	03	03		073	42	STO	
033	55	÷		074	07	07	
034	53	(		075	61	GTO	
035	93	.		076	13	C	
036	00	0					
037	04	4	r				
038	00	0					
039	02	2					
040	65	×					

Register Contents, Labels, and Data Cards--Algebraic System

Register	Contents	Labels	Contents
R_{00}	r	Label A	r
R_{01}	rpm	Label B	rpm
R_{02}	rps^2	Label C	R.C.F.
R_{03}	R.C.F.	Label D	t_1
R_{04}	t_1	Label E	t_2
R_{05}	$t_1 \times$ R.C.F.		
R_{06}	t_2		
R_{07}	rpm_2		

Example

A.   How many rpm are required to generate $100,000 \times g$, using the No. 40 rotor?

Solution

33,250 rpm

B.   When the No. 30 rotor is used at the maximum speed of 30,000 rpm, what is the relative centrifugal force?

Solution

$105,525 \times g$

C.   A standard sucrose density gradient centrifugation using the SW 50.1 rotor requires 12 hr at 35,000 rpm.   If you want to run for 10 hr instead, what rpm would be required to give you a sucrose density gradient equivalent to the standard run?

Solution

38,341 rpm

References

1.   Henry B. Bull (1964), *Introduction to Physical Biochemistry*, F.A. Davis Company, Philadelphia, Pa., pp. 20-21.

3B.   MOLECULAR WEIGHT FROM SEDIMENTATION VELOCITY DATA

An alternative approach to molecular weight determination
using the ultracentrifuge is the sedimentation velocity method.
Molecular sedimentation is described by the Svedberg equation:

$$S = \frac{1}{r\omega^2}\frac{dr}{dt} = \frac{1}{\omega^2}\frac{d\ln r}{dt} \tag{1}$$

where S = sedimentation coefficient (sec)
   r = distance of particles (protein molecules) from the axis
      of rotation
   t = time
   ω = angular velocity

In practice the centrifuge is brought to a constant rpm, and then
measurements of the distance of molecular movement in the centrifugal field (actually the average of the molecules sedimenting as
observed from the optical pattern) as a function of time are
taken.   These values are then plotted as $\ln r$ versus time, using
the calculator linear regression routine to obtain the value of
the slope of the line for computation of the observed sedimentation coefficient:

$$\text{Slope} = \omega^2 S_{obs} \tag{2}$$

Unlike the equilibrium centrifugation method, the sedimentation
velocity procedure requires other protein physical parameters
(viscosity, density, partial specific volume), which must be measured independently of the centrifugation experiment.   With this
additional information the sedimentation coefficient is corrected
to standard conditions (i.e., S for the protein in a solvent with
the density and viscosity of water at 20°C:

$$S^0_{20,w} = S^0_{obs}\frac{(1 - \bar{v}\rho)_{20,w}}{(1 - \bar{v}\rho)_{20,s}}\frac{\eta_{20,s}}{\eta_{20,w}} \tag{3}$$

where $\eta_{20,s}$ = viscosity of solution at 20°C
   $\eta_{20,w}$ = viscosity of water at 20°C
   $\bar{v}$ = partial specific volume (cm^3)
   $\rho_{20,w}$ = density of water at 20°C
   $\rho_{20,s}$ = density of solution at 20°C

The molecular weight may then be calculated utilizing the diffusion coefficient derived from diffusion measurements:

$$M.W. = \frac{S^0_{20,w}}{D^0_{20,w}} \frac{RT}{(1 - \bar{v}\rho)_{20,w}} \qquad (4)$$

User Instructions--RPN

Step	Instructions	Input	Keys	Output
1	Initialize: Clear Registers		f e	0.00
2	To set print flag		f d	1.00
3	To clear print flag		f d	0.00
4	Enter revolutions per minute	rpm	ENTER↑	rpm
5	Enter partial specific volume	$\bar{v}$	f a	
6	Enter density of solution at temperature of experiment	$\rho_{20,s}$	ENTER↑	$\rho_{20,s}$
7	Enter relative viscosity at temperature of experiment	rel.η	f b	rel.η
8	Enter diffusion coefficient of protein	$D_{20,w}$	ENTER↑	$D_{20,w}$
9	Enter temperature of experiment	T (°C)	f c	
10	Enter: time (min)	t	ENTER↑	t
	radius (cm)	r	A	n
	To delete incorrect data:			
11	Enter: time	t	ENTER↑	t
	radius	r	B	$n - 1$
12	Compute coefficient of determination and S value		C	r^2 S
13	To compute $S_{20,w}$		R/S	$S_{20,w}$
14	To compute molecular weight of protein		D	M.W.

Program Listing--RPN

Line	Key	Comments	Line	Key	Comments
001	*LBLe		045	ST07	$\to$ °K
002	CLRG	Initialize:	046	R↓	
003	P≠S	Clear Registers	047	ST06	$\to D_{20,w}$
004	CLRG		048	RTN	
005	CLX		049	*LBLA	
006	DSP3		050	LN	$\to \ln r$
007	FIX		051	X≠Y	$\to t$
008	RTN		052	F2?	
009	*LBLα		053	GSB9	Print if print flag
010	ST01	Store $\bar{v}$	054	Σ+	is set
011	.		055	RTN	Accumulate sums
012	9	Compute:	056	*LBLB	
013	9		057	LN	Delete incorrect
014	8	$(1 - \bar{v}\rho)_{20,w}$	058	X≠Y	data
015	x		059	F2?	
016	1		060	GSB3	
017	-		061	F2?	
018	CHS		062	GSB9	
019	ST05	$\to (1 - \bar{v}\rho)_{20,w}$	063	Σ-	
020	R↓		064	RTN	
021	2	Compute:	065	*LBLC	Compute coefficient
022	x		066	P≠S	of determination
023	Pi	$\omega = 2\pi \, (\text{rpm})/60$	067	RCL8	
024	x		068	RCL4	
025	6		069	RCL6	
026	0		070	x	
027	÷		071	RCL9	
028	ST00	$\to \omega$	072	÷	
029	RTN		073	-	
030	*LBLb		074	ENT↑	
031	ST03	$\to$ rel. η	075	ENT↑	
032	R↓		076	RCL4	
033	RCL1		077	X²	
034	x		078	RCL9	
035	1		079	÷	
036	-		080	RCL5	
037	CHS		081	X≠Y	
038	ST04	$\to (1 - \bar{v}\rho)_{20,s}$	082	-	
039	RTN		083	÷	
040	*LBLc		084	‾TOB	$\to$ slope
041	2		085	x	
042	7		086	RCL6	
043	3		087	X²	
044	+		088	RCL9	

Line	Key	Comments		Line	Key	Comments
089	÷			135	7	
090	CHS			136	RCL7	
091	RCL7			137	x	
092	+			138	RCL5	
093	÷			139	÷	
094	PRTX			140	x	
095	RCL6	→ r^2		141	DSP2	
096	RCL4			142	PRTX	Print molecular
097	RCLB			143	STOD	weight
098	x			144	RTN	
099	-			145	*LBLd	Print flag: Set and
100	RCL9			146	0	clear
101	÷			147	F2?	
102	STOA			148	RTN	
103	P⇄S			149	1	
104	RCL0	Compute S value		150	SF2	
105	X²			151	RTN	
106	6			152	*LBL9	Print subroutine
107	0			153	PRTX	
108	x			154	X⇄Y	
109	RCLB			155	PRTX	
110	X⇄Y			156	X⇄Y	
111	÷			157	SF2	
112	SCI			158	SPC	
113	DSP2			159	RTN	
114	PRTX	→ S		160	*LBL3	Print deletion
115	R/S			161	SPC	indicator sub-
116	RCL5	Corrected S value		162	DSP0	routine
117	RCL4			163	1	
118	÷			164	CHS	
119	RCL3			165	PRTX	
120	x			166	SF2	
121	x			167	DSP3	
122	PRTX	→ $S_{20,w}$		168	R↓	
123	STOC			169	RTN	
124	RTN					
125	*LBLD					
126	RCLC	Compute molecular				
127	RCL6	weight				
128	÷					
129	8					
130	.					
131	3					
132	1					
133	4					
134	EEX					

Register Contents, Labels, and Data Cards--RPN

Register	Contents	Labels	Contents
R_0	ω	A	$t \uparrow r(+)$
R_1	$\bar{\upsilon}$	b	$t \uparrow r(-)$
R_2	0	C	$\to r$, S; $S_{20,w}$
R_3	$\eta_{20,s}/\eta_{20,w}$	D	$\to$M.W.
R_4	$(1 - \bar{\upsilon}\rho)_{20,s}$	a	rpm $\uparrow \bar{\upsilon}$
R_5	$(1 - \bar{\upsilon}\rho)_{20,s}$	b	$\rho_{20,s} \uparrow$ rel. η
R_6	$D_{20,w}$	c	$D_{20,w} \uparrow T(^\circ C)$
R_7	$T(^\circ K)$	d	Print?
R_{s4}	$\sum t$	e	Initialize
R_{s5}	$\sum t^2$		
R_{s6}	$\sum \ln r$		
R_{s7}	$\sum (\ln r)^2$		
R_{s8}	$\sum t(\ln r)$		
R_{s9}	n		
R_A	Intercept		
R_B	Slope		
R_C	$S_{20,w}$		
R_D	M.W.		

User Instructions--Algebraic System

Step	Instructions	Input	Keys	Output
1	Enter revolutions per minute	rpm	A'	ω^2
2	Enter $\eta_{20,s}/\eta_{20,w}$	$\eta_{20,s}/\eta_{20,w}$	B'	
3	Enter $\bar{\upsilon}$	$\bar{\upsilon}$	C'	
4	Enter $\rho_{20,s}$	$\rho_{20,s}$	R/S	
5	Enter $D_{20,w}$	$D_{20,w}$	R/S	
6	Enter time	t	A	
7	Enter radius (cm)	r	B	$\ln r$
8	Enter all data pairs t,r before calculation of S_{obs}		R/S	n
9	Calculate S_{obs}		C	S_{obs}
10	Calculate correlation coefficient		R/S	r
11	Calculate $S_{20,w}$		D	$S_{20,w}.$
12	Calculate molecular weight		E	M.W.

Program Listing--Algebraic System

Line	Key	Entry	Comments	Line	Key	Entry	Comments
000	76	LBL		041	23	23	
001	16	A'	Enter rpm	042	91	R/S	
002	42	STO		043	42	STO	
003	11	11		044	15	15	
004	65	×		045	53	(	Enter $\rho_{20,s}$
005	93	.		046	01	1	
006	01	1		047	75	-	
007	00	0		048	53	(	
008	04	4		049	43	RCL	
009	07	7		050	14	14	
010	02	2		051	65	×	
011	95	=		052	43	RCL	
012	33	X²		053	15	15	
013	42	STO		054	54	)	
014	12	12		055	54	)	
015	91	R/S		056	95	=	
016	76	LBL		057	42	STO	
017	17	B'		058	16	16	
018	42	STO		059	91	R/S	
019	13	13		060	42	STO	
020	91	R/S	Enter	061	17	17	
021	76	LBL	$n_{20,s}/n_{20,w}$	062	91	R/S	
022	18	C'		063	76	LBL	Enter $D_{20,w}$
023	42	STO		064	11	A	
024	14	14	Enter $\bar{v}$	065	42	STO	
025	53	(		066	18	18	
026	01	1		067	76	LBL	
027	75	-		068	32	X⋮T	
028	53	(		069	36	PGM	
029	43	RCL		070	01	01	Enter time
030	14	14		071	32	X⋮T	
031	65	×		072	92	RTN	
032	93	.		073	91	R/S	
033	09	9		074	76	LBL	
034	09	9		075	12	B	
035	08	8		076	42	STO	
036	02	2		077	19	19	
037	54	)		078	95	=	
038	54	)		079	23	LNX	
039	95	=		080	91	R/S	Enter r
040	42	STO		081	78	Σ+	

Line	Key	Entry	Comments	Line	Key	Entry	Comments
082	92	RTN		123	43	RCL	
083	91	R/S		124	16	16	
084	76	LBL		125	54	)	
085	13	C		126	95	=	
086	69	OP		127	42	STO	
087	12	12		128	24	24	
088	32	X:T	n	129	91	R/S	
089	42	STO		130	76	LBL	
090	20	20	Call linear	131	15	E	
091	53	(	regression	132	53	(	
092	43	RCL		133	43	RCL	
093	20	20	s_{obs}	134	24	24	
094	55	÷		135	55	÷	
095	06	6		136	43	RCL	$s_{20,w}$
096	00	0		137	17	17	
097	54	)		138	54	)	
098	55	÷		139	65	×	
099	43	RCL		140	53	(	
100	12	12		141	02	2	
101	54	)		142	93	.	
102	95	=		143	04	4	
103	42	STO		144	04	4	
104	21	21		145	52	EE	
105	91	R/S		146	01	1	
106	69	OP		147	00	0	
107	13	13		148	55	÷	
108	42	STO		149	43	RCL	
109	22	22		150	23	23	
110	91	R/S		151	54	)	
111	76	LBL		152	95	=	
112	14	D	r	153	42	STO	
113	43	RCL		154	25	25	M.W.
114	13	13		155	91	R/S	
115	65	×					
116	43	RCL					
117	21	21					
118	65	×					
119	53	(					
120	43	RCL					
121	23	23					
122	55	÷					

Register Contents, Labels, and Data Cards--Algebraic System

Register	Contents	Labels	Contents
R0 → R6	Linear regression	Label A'	Enter rpm
R12	ω^2	Label B'	Enter $\eta_{20,s}/\eta_{20,w}$
R13	$\eta_{20,s}/\eta_{20,w}$	Label C'	Enter $\overline{\upsilon}$
R14	$\overline{\upsilon}$	Label A	Enter t
R16	$(1 - \overline{\upsilon}\rho)_{20,s}$	Label B	Enter r
R17	$D_{20,w}$	Label C	Calculate S_{obs}
R21	S_{obs}	Label D	Calculate $S_{20,w}$
R23	$(1 - \overline{\upsilon}\rho)_{20,s}$	Label E	Calculate M.W.
R24	$S_{20,w}$		

Example

A chromatographically pure soybean protein was run in a sedimentation velocity experiment. Determine the molecular weight from the experimental data given below.

Time (min)	Radius (cm)	ln r
0	6.11	1.81
10	6.20	1.83
20	6.29	1.84
30	6.37	1.85
40	6.46	1.87
50	6.55	1.88

rpm	$= 68,600$
$\eta_{20,s}/\eta_{20,w}$	$= 1.04$
$\overline{\upsilon}_{prot}$	$= 0.725$
ρ_{sol}	$= 1.002$
$D_{20,w}$	$= 3.49 \times 10^{-7}$ cm^2/sec

Solution

$$S_{obs} = 4.46 \times 10^{-13} \text{ sec}$$
$$r = 0.999$$
$$S^0_{20,w} = 4.69 \times 10^{-13} \text{ sec}$$
$$\text{M.W.} = 1.18 \times 10^5$$

References

1. C. Chervenka (1970), *A Manual of Methods for the Analytical Centrifuge*, Spinco Division of Beckman Instruments, Palo Alto, Calif.

3C. MOLECULAR WEIGHT BY EQUILIBRIUM SEDIMENTATION

Determination of the molecular weight of a macromolecule by equilibrium sedimentation is relatively straightforward in comparison to the velocity method. In the equilibrium method the analytical ultracentrifuge is run at a low constant speed (about 8000 rpm for a protein with a molecular weight of 60,000), and the molecular weight can be estimated using a single experimental method (i.e., separate determination of the diffusion coefficient is not necessary). However, both methods require knowledge of the partial specific volume of the macromolecule. To avoid a lengthy treatment of the theory of sedimentation equilibrium, we refer the reader to references 1 and 2 at the end of this section.

To compute molecular weight, the concentration of solute throughout the cell must be determined when the condition of equilibrium has been reached. In addition, the rotor speed, the temperature, and the density of the solution in the cell must be known accurately.

Some knowledge of two other important quantities is useful: these are the optimal run speed, and the time required to attain the experimental equilibrium condition--the *transient time*. If the approximate molecular weight of the solute under study is known, operating speeds and transient times can be estimated from equations available in references 3 and 4. The transient time can be reduced significantly by the technique of *overspeeding*. This is a simple procedure in which the rotor is initially run at a speed higher than the equilibrium speed selected. During this period, the bulk solute is transported by sedimentation toward the bottom of the cell, so that the quantities of material in the upper and lower halves of the cell at least grossly approximate those required for the final equilibrium condition. The rotor is then decelerated to a speed lower than the equilibrium speed to allow diffusion of the sharp gradients formed during the initial period. Finally, the rotor is accelerated to the equilibrium speed and allowed to run until the required solute distribution is attained. A detailed procedure for computing the required speeds and times of operation for each step (which requires a prior knowledge of molecular weight) has been described by Hexner, Radford, and Beams (5).

The concentration data required are most conveniently determined by means of the absorption optical system since the concentration data can be obtained directly from the absorbance of the cell contents. The development of the split-beam photoelectric scanning system has made the determination of molecular weights using absorption optics an easy, routine procedure. This method has the advantages that very low concentrations of solute can be

used, allowing more accurate extrapolations to infinite dilution,
and that the concentration data required are obtained directly
from the recorded data. A synthetic boundary run is not required,
and mass need not be conserved.

The calculator program provided in this section allows one
to enter optical density readings as a function of radial position. The optical density is read at a series of points across
the portion of the chart corresponding to the cell, using incre-
ments of radius such that 15-20 experimental points are obtained.
Optical densities (O.D.'s) need not be converted into corresponding values of concentration if the relationship between O.D. and
concentration is linear; if this is not the case and the correct
relationship is known, O.D. units are converted approximately into concentration before entering the data.

The equation for molecular weight used in this program is

$$M.W. = \frac{2RT}{(1 - \bar{\upsilon}\rho)\omega^2} \frac{d \ln c}{d(r^2)}$$

where M.W. = molecular weight
$\quad$ R $\quad$ = gas constant (8.314×10^7 ergs/degree·mole)
$\quad$ T $\quad$ = absolute temperature (°K)
$\quad$ $\bar{\upsilon}$ $\quad$ = partial specific volume (ml/gm)
$\quad$ ρ $\quad$ = density of solution (gm/ml)
$\quad$ ω $\quad$ = angular velocity (radians/sec)
$\quad$ c $\quad$ = concentration (O.D. units)
$\quad$ r $\quad$ = distance from axis of rotation (cm)

The quantity $\ln c$ is entered into the least squares program as
the dependent variable, and the r^2 as the independent variable.
The slope of the straight line is computed as the quantity
$d \ln c/d(r^2)$ and substituted in the equation for molecular weight.
The calculated molecular weight is rounded off and truncated to
the nearest 100 in this program, since for most low molecular
weight proteins the molecular weight as measured by the equilibrium sedimentation method is accurate only to within ±100 of the
actual value.

User Instructions--RPN

Step	Instructions	Input	Keys	Output
1	Initialize: Clear Registers		f a	0.00
2	Calculate magnification factor: enter distance between reference holes	cm	E	F
3	Input distance from rotor center to first reference hole (default value is 5.7 cm)	cm	f b	
4	Enter rpm and temperature (°C)	rpm T (°C)	ENTER↑ f c	
5	Enter density of solvent and partial specific volume of protein (Default value of $1 - \overline{v}\rho$ is 0.2671) $v = 0.73$ ml/gm $\rho = 1.004$ gm/ml	ρ $\overline{v}$	ENTER↑ f d	
6	Enter $r_{tracing}$: Calculate r_{true} using magnification factor and distance from rotor center	r_{trace}	A	r_{true}
7	Enter concentration (e.g., O.D. or mg/ml)	C	B	n
	To delete incorrectly entered data:			
8	Enter: r_{trace} C	r_{trace} C	A C	r_{true} $n - 1$
9	Repeat steps 6 and 7 for all pairs of data			

Step	Instructions	Input	Keys	Output
10	Compute coefficient of deter- mination and molecular weight (M.W. is rounded off to nearest 100)		D	r^2 M.W.
1a	To set data print flag		f e	1.00
1b	To clear data print flag		f e	0.00

Program Listing--RPN

Line	Key	Comments	Line	Key	Comments
001	*LBLa		033	x	
002	CLRG	Initialize:	034	6	
003	P⇄S	Clear registers and	035	0	
004	CLRG	store default values	036	÷	
005	FIX		037	X²	
006	DSP2		038	STOE	
007	5		039	RTN	$\to \omega^2$
008	.		040	*LBLd	
009	7	Store distance to	041	x	
010	STO1	rotor center	042	1	Compute and store
011	.		043	X⇄Y	$(1 - \overline{\upsilon}\rho)$
012	2		044	-	
013	6		045	STOC	
014	7		046	CLX	
015	1	$\to (1 - \overline{\upsilon}\rho)$	047	RTN	
016	STOC		048	*LBLE	
017	CLX		049	1	Compute magnifica-
018	RTN		050	.	tion factor
019	*LBLb	Store distance to	051	6	
020	STO1	rotor center	052	1	
021	CLX		053	÷	
022	RTN		054	STO0	$\to F$
023	*LBLc		055	RTN	
024	2		056	*LBLA	
025	7		057	RCL0	
026	3		058	÷	Compute r_{true}
027	+	$\to T(^\circ K)$	059	RCL1	
028	STOD		060	+	
029	R↓		061	RTN	
030	2		062	*LBLB	Enter r_{true} and
031	Pi		063	LN	concentration
032	x		064	X⇄Y	

Line	Key	Comments	Line	Key	Comments
065	x^2		111	RCL6	
066	F2?		112	RCL4	
067	GSB9		113	RCLB	
068	$\Sigma+$	Data summation	114	x	
069	RTN		115	-	
070	*LBLC	Data deletion	116	RCL9	
071	LN	routine	117	$\div$	
072	X$\rightleftarrows$Y		118	STOA	$\rightarrow$ intercept
073	x^2		119	8	
074	F2?		120	.	
075	GSB3		121	3	
076	F2?		122	1	
077	GSB9		123	4	
078	$\Sigma-$		124	EEX	
079	RTN		125	7	
080	*LBLD		126	RCLD	
081	P$\rightleftarrows$S		127	x	
082	SPC		128	2	
083	RCL8		129	x	Compute:
084	RCL4		130	RCLC	$2RT/(1 - \bar{\upsilon}\rho)\omega^2$
085	RCL6	Compute coefficient	131	RCLE	
086	x	of determination	132	x	
087	RCL9	and molecular weight	133	$\div$	
088	$\div$		134	RCLB	Slope X
089	-		135	x	$[2RT/(1 - \bar{\upsilon}\rho)\omega^2]$
090	ENT↑		136	SCI	
091	ENT↑		137	DSP2	
092	RCL4		138	RND	
093	x^2		139	FIX	
094	RCL9		140	P$\rightleftarrows$S	
095	$\div$		141	PRTX	Round off and print
096	RCL5		142	RTN	M.W. to nearest 100
097	X$\rightleftarrows$Y		143	*LBL3	
098	-		144	SPC	
099	$\div$		145	DSP0	Print deletion
100	STOB	$\rightarrow$ slope	146	1	indicator
101	x		147	CHS	
102	RCL6		148	PRTX	
103	x^2		149	DSP2	
104	RCL9		150	R↓	
105	$\div$		151	SF2	
106	CHS		152	RTN	
107	RCL7		153	*LBL9	
108	+		154	PRTX	Print subroutine
109	$\div$		155	X$\rightleftarrows$Y	
110	PRTX	$\rightarrow$ r^2			

Line	Key	Comments
156	PRTX	
157	X≠Y	
158	SPC	
159	SF2	
160	RTN	
161	*LBLe	Print flag:
162	0	Set and Clear
163	F2?	
164	RTN	
165	1	
166	SF2	
167	RTN	

Register Contents, Labels, and Data Cards--RPN

Register	Contents	Register	Contents	Labels	Contents
R_0	F			A	$r_{tracing}$
R_1	cm to center			B	$r_{true} \uparrow C(+)$
R_{s4}	$\sum x$			C	$r_{true} \uparrow C(-)$
R_{s5}	$\sum x^2$			D	$\rightarrow r^2$, M.W.
R_{s6}	$\sum y$			E	$F \uparrow$
R_{s7}	$\sum y^2$			a	Initialize
R_{s8}	$\sum xy$			b	cm to center
R_{s9}	n			c	rpm $\uparrow T$ (°C)
R_C	$(1 - \bar{v}\rho)$			d	$\rho \uparrow \bar{v}$
R_D	T (°K)			e	Print?
R_E	ω^2				

User Instructions--Algebraic System

Step	Instructions	Input	Keys	Output
1	Print toggle		2nd E'	
2	Enter partial specific volume of protein	$\bar{v}$	2nd A'	
3	Enter distance between reference holes; calculate magnification factor	cm	R/S	F
4	Enter rpm	rpm	2nd B'	
5	Enter distance from rotor center to first reference hole (default entry is 5.7 cm)	cm	R/S	

Step	Instructions	Input	Keys	Output
6	Enter T (°C)	T(°C)	2nd C'	
7	Enter $r_{tracing}$: Calculate r_{true} using magnification factor and distance from center	r_{trace}	A	r^2_{true}
8	Enter r^2_{true} into linear regression		R/S	
9	Enter concentration	C	B	ln C
10	Enter ln C into linear regression; enter all data pairs before molecular weight calculation	ln C	R/S	n
11	Determine molecular weight		C	M.W.
12.	Determine correlation coefficient for fitted line		D	r

Program Listing--Algebraic System

Line	Key	Entry	Comments	Line	Key	Entry	Comments
000	76	LBL		017	29	29	
001	10	E'	Print toggle	018	02	2	
002	87	IFF		019	00	0	
003	08	08		020	71	SBR	Call print
004	45	Y×		021	85	+	subroutine
005	86	STF		022	43	RCL	
006	08	08		023	29	29	
007	92	RTN		024	75	-	
008	76	LBL		025	01	1	
009	45	Y×		026	95	=	
010	22	INV		027	94	+/-	
011	86	STF		028	42	STO	$(1 - \bar{v}\rho)$
012	08	08		029	11	11	
013	92	RTN		030	91	R/S	
014	76	LBL $\bar{v}\rho$		031	42	STO	Compute mag-
015	16	A'		032	12	12	nification
016	42	STO		033	55	÷	factor

Line	Key	Entry	Comments	Line	Key	Entry	Comments
034	01	1		078	43	RCL	
035	93	.		079	29	29	
036	06	6		080	85	+	
037	01	1		081	02	2	
038	95	=		082	07	7	
039	42	STO		083	03	3	
040	12	12	$\to F$	084	95	=	T (°K)
041	91	R/S		085	65	×	
042	76	LBL		086	08	8	
043	17	B'	Enter rpm	087	93	.	
044	42	STO		088	03	3	
045	29	29		089	01	1	
046	02	2		090	03	3	R
047	01	1		091	52	EE	
048	71	SBR	Call print	092	07	7	
049	85	+	subroutine	093	55	÷	
050	43	RCL		094	53	(	RT
051	29	29		095	43	RCL	
052	65	×		096	11	11	
053	93	.		097	65	×	
054	01	1		098	43	RCL	
055	00	0	$\omega = 2\pi\text{rpm}/60$	099	16	16	
056	04	4		100	95	=	
057	07	7		101	42	STO	$RT(1 - \bar{v}\rho)\omega^2$
058	02	2		102	19	19	
059	95	=		103	22	INV	
060	33	X²	ω^2	104	52	EE	
061	42	STO		105	91	R/S	
062	16	16		106	76	LBL	
063	05	5		107	11	A	
064	93	.		108	42	STO	r_{trace}
065	07	7		109	29	29	
066	91	R/S		110	55	÷	
067	42	STO	Store distance	111	43	RCL	
068	13	13	to rotor cen-	112	12	12	
069	91	R/S	ter	113	54	)	
070	76	LBL	T(°C)	114	85	+	
071	18	C'		115	43	RCL	
072	42	STO		116	13	13	
073	29	29		117	95	=	
074	02	2		118	42	STO	
075	02	2		119	29	29	
076	71	SBR	Call print	120	02	2	
077	85	+	subroutine	121	03	3	

Line	Key	Entry	Comments	Line	Key	Entry	Comments
122	71	SBR	Call print	166	32	X:T	
123	85	+	subroutine	167	65	×	
124	43	RCL		168	43	RCL	
125	29	29		169	19	19	
126	33	X²	r^2_{true}	170	65	×	
127	42	STD		171	02	2	
128	29	29		172	95	=	
129	02	2		173	42	STD	
130	04	4		174	29	29	
131	71	SBR	Call print	175	02	2	
132	85	+	subroutine	176	07	7	
133	43	RCL		177	71	SBR	Call print
134	29	29		178	85	+	subroutine
135	91	R/S		179	91	R/S	
136	36	PGM	linear	180	76	LBL	
137	01	01	regression	181	14	D	r
138	32	X:T		182	69	DP	
139	91	R/S		183	13	13	
140	76	LBL		184	42	STD	
141	12	B	C	185	29	29	
142	42	STD		186	02	2	
143	29	29		187	08	8	
144	02	2		188	71	SBR	Call print
145	05	5		189	85	+	subroutine
146	71	SBR	Call print	190	91	R/S	
147	85	+	subroutine	191	76	LBL	
148	43	RCL		192	85	+	
149	29	29		193	87	IFF	
150	23	LNX	ln C	194	08	08	
151	42	STD		195	95	=	Print
152	29	29		196	43	RCL	subroutine
153	02	2		197	29	29	
154	06	6		198	92	RTN	
155	71	SBR	Call print	199	76	LBL	
156	85	+	subroutine	200	95	=	
157	91	R/S		201	42	STD	
158	43	RCL		202	07	07	
159	29	29		203	73	RC*	
160	78	Σ+	linear	204	07	07	
161	91	R/S	regression	205	69	DP	
162	76	LBL	M.W.	206	04	04	
163	13	C		207	43	RCL	
164	69	DP		208	29	29	
165	12	12		209	69	DP	

Line	Key	Entry	Comments
210	06	06	
211	69	□P	
212	00	00	
213	98	ADV	
214	92	RTN	

Register Contents, Labels, and Data Cards--Algebraic System

Register		Contents	Labels	Contents
R0	R6	Linear regression	Label A'	Enter $\bar{\upsilon}$
R11		$(1 - \bar{\upsilon}\rho)$	Label B'	Enter rpm
R16		ω^2	Label C'	Enter $T(°C)$
R19		$RT/(1 - \bar{\upsilon}\rho)\omega^2$	Label A	Enter r
			Label B	Enter C
			Label C	Derive M.W.
			Label D	Derive r
			Label E'	Print toggle

Data Card (4 2nd write)

Alpha Print	Code Register	Contents
42330000.	20	VP
35333000.	21	RPM
37173033.	22	TEMP
35000000.	23	R
35700000.	24	R²
15323115.	25	CONC
27312015.	26	LN-C
30430000.	27	MW
15323535.	28	CORR

Example

Compute the molecular weight of the protein from the sedimentation equilibrium data given in Table 1.   Distance between reference holes is 18.7 cm.   Figure 3C shows the plot of these data in the form necessary to compute the molecular weight ($\ln C$ vs. r^2).

TABLE 1

r_{trace} (cm)	r_{true} (cm)	Concentration
12.89	6.81	3.06
13.13	6.83	3.54
13.71	6.88	4.55
14.17	6.92	5.53
14.52	6.95	6.50
14.87	6.98	7.50
15.10	7.00	8.51
15.45	7.03	9.54
15.68	7.05	10.51
15.80	7.06	11.50
16.03	7.08	12.50
16.15	7.09	13.43
16.38	7.11	14.21

$$\bar{\upsilon} = 0.723$$
$$\text{rpm} = 20,000$$
$$\text{Temperature} = 10°C$$

Solution

r^2	$\ln C$
46.37	1.18
46.65	1.26
47.33	1.52
47.89	1.71
48.30	1.87
48.72	2.02
49.00	2.14

r^2	ln C
49.42	2.26
49.70	2.35
49.84	2.44
50.13	2.53
50.27	2.61
50.55	2.65

M.W. = 1.422×10^4
r = 0.999

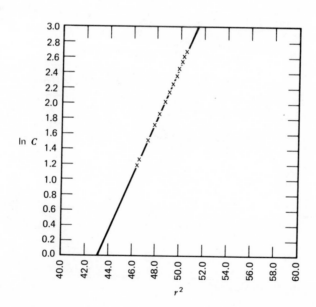

Figure 3C. Molecular weight by equilibrium sedimentation

References

1. Instruction Manual, Beckman Model E Analytical Ultracentrifuge.

2. H.K. Schachman, "Ultracentrifugation, Diffusion and Viscosity," *Methods in Enzymology IV,* 32-103.

3. K.E. Van Holde (1967), "Sedimentation Equilibrium," *Fractions
   1,* 1.

4. T. Svedberg and K.O. Pedersen (1940), *The Ultracentrifuge,*
   Clarendon Press, Oxford, p. 16.

5. P.E. Hexner, L.E. Radford, and J.W. Beams (1969), "Achieve-
   ment of Sedimentation," *Proceedings of the National Academy
   of Science 47,* 1848.

6. A. Katchalsky and P.F. Curran (1967), *Nonequilibrium Thermo-
   dynamics in Biophysics,* Harvard University Press, Cambridge,
   Mass., pp. 104-112.

IV
LIGAND BINDING
AND KINETICS

4A. SCATCHARD PLOT FOR EQUILIBRIUM BINDING DATA

The study of ligand binding to macromolecules plays a significant role in biochemical and biomedical research. The collection and the analysis of equilibrium binding data are central to studies in enzyme kinetics, immunoassay, and pharmacology. Many proteins bind small molecules but do not catalyze a reaction of the ligand. Some examples include the binding of oxygen by myoglobin or hemoglobin and the binding of hormones and drugs by specific receptor proteins. The binding of substrates, inhibitors, and activators to enzymes can also be studied by equilibrium methods if no catalytic reaction occurs. The typical experimental design involves a labeled ligand and a method for separating the ligand bound to a macromolecule from the free ligand. Knowledge of the total ligand and the total protein and measurement of the bound ligand permit determination of the bound fraction and free concentration of the ligand. Equilibrium binding data are usually analyzed by a Scatchard plot.

The generalized Scatchard model is given by the following equation

$$\overline{v} = \frac{N_1 K_1 A}{1 + K_1 A} + \cdots + \frac{N_i K_i A}{1 + K_i A} \tag{1}$$

where $\overline{v}$ = total number of moles of bound A per total protein concentration

N_i = number of sites of class i

K_i = site-specific binding (association) constant

147

This model assumes that all N_i sites are identical but completely independent. Since $K_{association} = 1/K_{dissociation}$, equation 1 can be expressed alternatively in terms of dissociation constants:

$$\bar{\nu} = \frac{N_1 A}{Kd_1 + A} + \ldots + \frac{N_i A}{Kd_i + A} \tag{2}$$

A second model is the stepwise equilibrium model, which is given by the following equation:

$$\bar{\nu} = \frac{K_1 A + 2K_1 K_2 A^2}{1 + K_1 A + K_1 K_2 A} \tag{3}$$

for a protein with two ligand binding sites. This more general formulation of the stepwise equilibrium model contains the generalized Scatchard model described by equation 1 as a special case.

A linearized form of the general Scatchard model is as follows:

$$\frac{\bar{\nu}}{A} = -\frac{1}{K_d}(\bar{\nu}) + \frac{N}{K_d} \tag{4}$$

A plot of $\bar{\nu}/A$ (i.e., moles of ligand bound per mole of enzyme divided by the concentration of free ligand) versus $\bar{\nu}$ is linear with a slope of $-1/K_d$ (Figure 4A). The intercept on the vertical axis gives N/K_d. The intercept on the horizontal axis gives N, the number of ligand binding sites per molecule of protein.

If the protein possesses multiple independent binding sites with different affinities for the ligand, the plot will be curved. There are, however, other possible explanations for a curvature of a Scatchard plot, and caution should be exercised in the interpretation of this type of analysis. For a very good analysis of the sources of error in equilibrium binding data see the paper by Cornish-Bowden and Koshland (6). This paper is specifically related to Hill plots but is applicable in general to all forms of equilibrium binding data.

This calculator program computes r^2, the coefficient of determination, K_d, the dissociation constant, and N, the total number of binding sites, by the linear regression method. Provision is made for outputting K_d, in engineering notation. By selecting the print option, the values of $\bar{\nu}/A$ and $\bar{\nu}$ will be computed and printed on the HP-97 before entering the summation registers. This allows for easier plotting of the reduced data. The best fit curve may be computed for plotting purposes as well.

User Instructions--RPN

Step	Instructions	Input	Keys	Output
1	Initialize: Clear Registers		f a	0.00
2	To set print flag		f e	1.00
3	To clear print flag		f e	0.00
4	To set engineering notation flag: this results in formatting $\overline{v}/A$ and K_d in engineering notation		f d	1.00
5	To clear engineering notation flag		f d	0.00
6	Enter: ligand concentration molar binding ratio	A $\overline{v}$	ENTER↑ A	A n
	To delete incorrect data:			
7	Enter: ligand concentration molar binding ratio	A $\overline{v}$	ENTER↑ B	A $n - 1$
	Repeat step 6 for all data			
8	Compute coefficient of determination and K_d		C	r^2 K_d
9	Compute number of ligand binding sites, N		D	N
	To compute best fit straight line:			
10	Enter molar binding ratio	$\overline{v}$	E	$\overline{v}/A$

Program Listing--RPN

Line	Key	Comments	Line	Key	Comments
001	*LBLa		045	-	
002	CLRG	Initialize: Clear	046	÷	
003	P⇄S	Registers	047	STOB	
004	CLRG		048	x	
005	RTN		049	RCL6	
006	*LBLA		050	X²	
007	÷	Enter $\bar{v}$	051	RCL9	
008	LSTX	Enter A	052	÷	Coefficient of deter-
009	X⇄Y		053	CHS	mination
010	1/X	→ $\bar{v}/A$	054	RCL7	
011	X⇄Y		055	+	
012	F2?		056	÷	→ r^2
013	GSB9	Print data if print	057	PRTX	
014	Σ+	flag is set	058	RCL6	
015	RTN	Accumulate sums	059	RCL4	
016	*LBLB		060	RCLB	
017	÷		061	x	
018	LSTX	Deletion of data	062	-	
019	X⇄Y		063	RCL9	
020	1/X		064	÷	
021	X⇄Y		065	STOA	
022	F2?	Print deletion	066	RCLB	
023	GSB3	indicator if print	067	CHS	Dissociation constant
024	F2?	flag is set	068	1/X	
025	GSB9	Delete	069	F0?	
026	Σ-		070	ENG	→ K_d
027	RTN		071	PRTX	
028	*LBLC		072	SPC	
029	P⇄S		073	P⇄S	
030	RCL8		074	RTN	
031	RCL4		075	*LBLD	
032	RCL6	Compute regression	076	RCLA	
033	x	coefficients	077	RCLB	Number of ligand
034	RCL9		078	CHS	binding sites
035	÷		079	÷	
036	-		080	FIX	→ N
037	ENT↑		081	PRTX	
038	ENT↑		082	RTN	
039	RCL4		083	*LBLE	
040	X²		084	RCLB	
041	RCL9		085	X⇄Y	Plotting/projection
042	÷		086	x	using best fit param-
043	RCL5		087	LSTX	eters
044	X⇄Y		088	X⇄Y	

Line	Key	Comments
089	RCLA	
090	+	
091	F2?	
092	GSB8	
093	RTN	
094	*LBLe	
095	0	Print flag:
096	F2?	Set and Clear
097	RTN	
098	1	
099	SF2	
100	RTN	
101	*LBL9	
102	X≷Y	Print subroutine
103	F0?	
104	ENG	
105	PRTX	
106	X≷Y	
107	FIX	
108	PRTX	
109	SF2	
110	SPC	
111	RTN	
112	*LBL3	
113	SPC	
114	DSP0	Deletion flag
115	1	indicator
116	CHS	
117	PRTX	
118	SF2	
119	DSP3	
120	R↓	
121	RTN	
122	*LBLd	
123	0	
124	F0?	Engineering notation
125	GTO0	flag:
126	SF0	Set and Clear
127	1	
128	RTN	
129	*LBL0	
130	CF0	
131	RTN	
132	*LBL8	
133	X≷Y	
134	FIX	

Line	Key	Comments
135	PRTX	
136	X≷Y	
137	F0?	Print:
138	ENG	→ $\bar{v}$
139	PRTX	→ $\bar{v}/A$
140	SF2	
141	RTN	

Register Contents, Labels, and Data Cards--PRN

Register	Contents	Labels	Contents
R_{s4}	$\sum \bar{v}$	A	$\bar{v} \uparrow A(+)$
R_{s5}	$\sum \bar{v}^2$	B	$\bar{v} \uparrow A(-)$
R_{s6}	$\sum \bar{v}/A$	C	$\rightarrow r^2, K_d$
R_{s7}	$\sum (\bar{v}/A)^2$	D	$\rightarrow N$
R_{s8}	$\sum (\bar{v}/A)(\bar{v})$	E	$\bar{v} \rightarrow \bar{v}/A$
R_{s9}	n	a	Initialize
R_A	N/K_d	d	Eng.?
R_B	$-1/K_d$	e	Print?

User Instructions--Algebraic System

Step	Instructions	Input	Keys	Output
1	Engineering notation flag set option: this results in $\overline{v}/A$ and K_d in engineering notation format		2nd A'	
2	Enter ligand concentration	A	A	
3	Enter molar binding ratio and derive $\overline{v}/A$	$\overline{v}$	B	$\overline{v}/A$
4	Enter $\overline{v}/A$ into linear regression; enter all data pairs before K_d calculation	$\overline{v}/A$	R/S	n
5	Calculate dissociation constant		C	K_d
6	Determine correlation coefficient		R/S	r
7	Calculate number of ligand binding sites		D	N
8	Enter molar binding ratio derive $\overline{v}/A$	$\overline{v}$	E	$\overline{v}/A$

Program Listing--Algebraic System

Line	Key	Entry	Comments	Line	Key	Entry	Comments
000	76	LBL		011	86	STF	
001	17	B'	Print toggle	012	08	08	
002	87	IFF		013	92	RTN	
003	08	08		014	76	LBL	Set flag for
004	45	YX		015	16	A'	engineering
005	86	STF		016	86	STF	notation
006	08	08		017	00	00	
007	92	RTN		018	91	R/S	
008	76	LBL		019	76	LBL	Enter A
009	45	YX		020	11	A	
010	22	INV		021	42	STO	

Line	Key	Entry	Comments	Line	Key	Entry	Comments
022	10	10		065	85	+	
023	42	STO	Call print	066	91	R/S	Linear
024	29	29	subroutine	067	43	RCL	regression
025	01	1		068	29	29	
026	04	4		069	78	Σ+	
027	71	SBR		070	91	R/S	Calculate K_d
028	85	+		071	76	LBL	
029	22	INV		072	13	C	
030	57	ENG	Enter	073	69	OP	
031	91	R/S		074	12	12	
032	76	LBL		075	32	X:T	
033	12	B		076	35	1/X	
034	42	STO		077	95	=	
035	11	11		078	94	+/-	
036	42	STO	Call print	079	42	STO	
037	29	29	subroutine	080	13	13	Call print
038	01	1		081	42	STO	subroutine
039	05	5		082	29	29	
040	71	SBR		083	01	1	
041	85	+	Call linear	084	08	8	
042	43	RCL	regression	085	71	SBR	
043	29	29	subroutine	086	85	+	
044	36	PGM		087	43	RCL	
045	01	01		088	29	29	
046	32	X:T		089	91	R/S	
047	43	RCL		090	69	OP	
048	11	11		091	13	13	
049	55	÷		092	42	STO	
050	43	RCL		093	29	29	r
051	10	10	$\bar{v}/A$	094	01	1	
052	95	=		095	09	9	
053	42	STO	Flag for	096	71	SBR	
054	12	12	engineering	097	85	+	Call print
055	87	IFF	notation	098	43	RCL	subroutine
056	00	00		099	29	29	
057	39	COS		100	91	R/S	
058	43	RCL	Call print	101	76	LBL	
059	12	12	subroutine	102	14	D	
060	42	STO		103	69	OP	
061	29	29		104	12	12	
062	01	1		105	42	STO	Calculate N
063	06	6		106	14	14	
064	71	SBR		107	65	×	

Line	Key	Entry	Comments	Line	Key	Entry	Comments
108	43	RCL		151	76	LBL	
109	13	13		152	38	SIN	
110	95	=		153	22	INV	
111	42	STD		154	86	STF	
112	29	29		155	00	00	
113	02	2		156	22	INV	
114	00	0		157	57	ENG	
115	71	SBR	Call print	158	91	R/S	
116	85	+	subroutine	159	76	LBL	Print
117	43	RCL		160	85	+	subroutine
118	29	29		161	87	IFF	
119	91	R/S		162	08	08	
120	76	LBL	Enter $\bar{\nu}$	163	95	=	
121	15	E		164	43	RCL	
122	42	STD		165	29	29	
123	29	29		166	92	RTN	
124	02	2		167	76	LBL	
125	01	1		168	95	=	
126	71	SBR	Call print	169	42	STD	
127	85	+	subroutine	170	25	25	
128	43	RCL		171	73	RC*	
129	29	29		172	25	25	
130	69	DP	Derive $\bar{\nu}$/A	173	69	DP	
131	14	14		174	04	04	
132	42	STD		175	43	RCL	
133	29	29		176	29	29	
134	02	2		177	69	DP	
135	02	2		178	06	06	
136	71	SBR	Call print	179	69	DP	
137	85	+	subroutine	180	00	00	
138	91	R/S		181	98	ADV	
139	43	RCL		182	92	RTN	
140	29	29					
141	76	LBL	Subroutine				
142	39	CDS	for engineer-				
143	43	RCL	ing notation				
144	12	12					
145	57	ENG					
146	42	STD					
147	12	12					
148	61	GTD					
149	00	00					
150	56	56					

Register Contents, Labels, and Data Cards--Algebraic System

Register	Contents	Labels	Contents
R0 → R6	Linear regression	Label A'	$St\ Flag\ 0$
R10	A	Label B'	print toggle
R11	$\bar{\nu}$	Label A	Enter A
R12	$\bar{\nu}/A$	Label B	Enter $\bar{\nu}$
R13	K_d	Label C	K_d
		Label D	N
		Label E	$\bar{\nu}/A'$

Data Card (4 2nd Write)

Alpha	Print Code	Register	Contents
15323115.		14	CONC
30401440.		15	M.B.
42631300.		16	V/A
26160000.		18	KD
15323535.		19	CORR
14243116.		20	BIND
42000000.		21	V
42631365.		22	V/A'

Example

The data given below were extracted from a paper by Cantley and Hammes (7) dealing with the binding of N^6-ethenoadenylylimidotriphosphate (εAMP-PNP), a fluorescent, nonhydrolyzable analog of adenosine triphosphate (ATP), by the chloroplast coupling factor adenosine triphosphatase of photosynthetic phosphorylation. Compute the dissociation constant from these data and the number of binding sites, N. Figure 4A shows the data plotted as $\overline{\nu}/A$ vs. $\overline{\nu}$.

TABLE 1

$\overline{\nu}$	A (μM)
0.03	0.53
0.09	1.48
0.16	2.67
0.24	4.14
0.30	5.24
0.39	6.72
0.63	14.09
0.82	21.48
1.06	30.17
1.30	44.69
1.54	76.68
1.52	126.67

Solution

$$r = 0.976$$
$$r^2 = 0.952$$
$$K_d = 33.719$$
$$n = 2.162$$

Best fit straight line

$\bar{\nu}$	$\bar{\nu}/A$
0	0.064
0.10	0.061
0.20	0.058
0.40	0.052
0.80	0.040
1.20	0.029
1.60	0.017

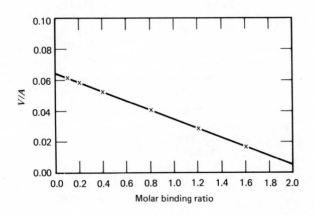

Figure 4A.   Scatchard plot--equilibrium binding data

References

1.   John E. Fletcher (1977), "A Generalized Approach to Equili-
     brium Models," *Journal of Physical Chemistry 81*, 2374-2378.

2.   I.H. Segel (1976), *Biochemical Calculations, 2nd Ed.*, John
     Wiley & Sons, Inc., New York, pp. 241-243.

3.   I.H. Segel (1975), *Enzyme Kinetics--Behavior and Analysis of
     Rapid Equilibrium and Steady State Enzyme Systems*, John Wiley
     & Sons, Inc., New York, pp. 218-220.

4.   Finn Wold (1971), *Macromolecules: Structure and Function*,
     Prentice-Hall, Inc., Englewood Cliffs, New Jersey, pp. 21-29.

5.   G. Scatchard (1949), *Annals   of the New York Academy of
     Science 51*, 660.

6.   A. Cornish-Bowden and D.E. Koshland, Jr. (1975), Diagnostic
     Uses of the Hill (Logit and Nernst) Plots," *Journal of
     Molecular Biology 95*, 201-212.

7.   L.C. Cantley, Jr. and G.G. Hammes (1975), "Characterization
     of Nucleotide Binding Sites on Chloroplast Coupling Factor
     1," *Biochemistry 14*, 2968-2975.

4B. ALLOSTERIC ENZYME KINETICS

Many enzyme reactions do not follow simple Michaelis-Menten kinetic behavior (i.e., hyperbolic saturation kinetics). This is particularly true for large multisubunit proteins with several substrate binding sites. In these types of regulatory or "allosteric" enzymes, the binding of one substrate or ligand can induce a structural or electronic change in the protein and thereby influence the binding of additional substrate molecules to adjacent sites on the enzyme. In some cases substrate or ligand binding to a regulatory protein can enhance substrate binding (positively cooperative kinetics) while in other instances it can diminish substrate-enzyme interactions (negatively cooperative kinetics) depending on the intrinsic nature of the regulatory protein.

Some preliminary indications of cooperative kinetic behavior is afforded by deviations from linearity of the Lineweaver-Burk plot. When the double reciprocal plot displays a downward concave trend, negative cooperativity is implied while concave upward behavior suggests positive cooperativity. However, particular caution should be used in interpreting the nonlinear behavior of double reciprocal plots as an indication of cooperative kinetics since factors such as protein denaturation or the requirement for the formation of an ordered ternary complex can produce similar curvilinear behavior.

For many proteins it has been convenient to express their cooperative kinetic properties by equation 1:

$$\frac{v}{V_{max}} = \frac{S^n}{K' + S^n} \tag{1}$$

where n = the number of substrate binding sites per molecule of
　　　enzyme (Hill coefficient)

　　　K' = a constant comprising the site interaction factors and
　　　the intrinsic dissociation constant, K_S

The above equation is known as the Hill equation (reference 4).

The constant K' in equation 1 no longer equals the substrate concentration that yields half-maximal velocity except when $n = 1$, when the equation reduces to the Michaelis-Menten equation.

If experimental velocity data are analyzed in terms of the Hill equation, the calculated value of n will almost always be less than the actual number of sites. The next highest integer above this *apparent* n value represents the minimum number of actual sites. Therefore, if the experimental data yields an n_{app} value of 1.8 based on the Hill equation, in effect the enzyme behaves as if it possesses exactly 1.8 substrate binding sites with very strong positive cooperativity.

The numerical value of the Hill coefficient also gives information on the type of cooperativity. For example, a Hill plot with $n > 1$ indicates positive cooperativity, $n = 1$ suggests noncooperative kinetics (i.e., Michaelis-Menten kinetics), and $n < 1$ suggests negative cooperativity.

The Hill equation can be converted to a useful linear form:

$$\log\left[\frac{v}{V_{max} - v}\right] = n\log S - \log K' \tag{2}$$

Thus, a plot of $\log [v/(V_{max} - v)]$ versus $\log S$ should yield a straight line with a slope of n. When $\log [v/(V_{max} - v)] = 0$, $v/(V_{max} - v) = 1$ and the corresponding position on the log S axis gives $S_{0.5}$. K' may be calculated from the relationship $K' = \sqrt[n]{S_{0.5}}$. Theoretically, the Hill plot is linear over the entire range of substrate concentration by virtue of the assumption that intermediate binding of ligand or substrate is not observed. However, this is true only for the special case in which all the dissociation constants are equal and the Hill coefficient $n = 1$. Deviations from linearity may be attributed to differences in dissociation constants and to the fact that at low specific velocities, complexes containing less than n molecules of substrate contribute significantly to the initial velocity. The limiting slopes (asymptotes) at very low and very high substrate concentrations are 1.0. It is often useful to sketch in the asymptotes at both extremes of substrate concentration to gain maximum information from a Hill plot (see reference 3).

Systematic error is likely to create severe problems whether one uses equilibrium binding data or velocity data in the analysis of Hill plots. Equilibrium binding data requires an accurate knowledge of protein concentration and the number of binding sites per molecule. Velocity data requires an accurate knowledge of V_{max}. The effect of an error in the estimation of V_{max} is to raise or lower the fractional saturation (v/V_{max}) by a constant percentage. Since $\log [v/(V_{max} - v)]$ is not proportional to v/V_{max}, the resulting error in the ordinate as plotted is not a constant percentage, but is relatively much larger at larger values. This can distort the true shape of the plot very considerably.

A thorough derivation of the Hill equation as well as a practical guide to the analysis and usefulness of Hill plots can be found in the references.

These calculator programs have been designed to accept kinetic data as inputs. This is valid if v/V_{max} is proportional to the fraction of the total number of binding sites occupied by the ligand. This is frequently true but it does not have to be (5). Therefore, deductions reached by analysis of Hill plots using kinetic data must be considered working hypotheses until equilibrium binding data have been obtained.

The calculator accepts V_{max}, v, and S and computes $\log [v/(V_{max} - v)]$ and $\log S$ for plotting. These values are then entered into the linear regression subroutine which yields the slope of the line n, the correlation coefficient r (coefficient of determination r^2 for the HP-67), the constant K' from the X-intercept, and the $S_{0.5}$ value derived from K'. Figure 4B shows a representative Hill plot for allosteric enzyme kinetics.

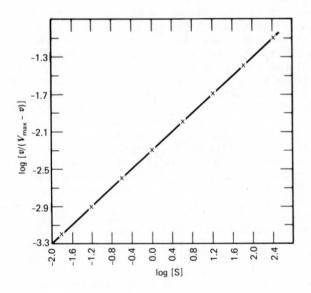

Figure 4B. Hill plot--allosteric enzyme kinetics

User Instructions--RPN

Step	Instructions	Input	Keys	Output
1	Initialize: Clear Registers		f e	0.00
2	To set print flag		f d	1.00
3	To clear print flag		f d	0.00
4	Enter V_{max}	V_{max}	f a	V_{max}
5	Enter: substrate concentration	S	ENTER↑	S
	velocity	v	A	n
	To delete incorrect data:			
6	Enter: substrate concentration	S	ENTER↑	S
	velocity	v	B	$n - 1$
	Repeat step 5 for all data			
7	Compute coefficient of determination, apparent Hill coefficient, and K'		C	r^2 n_{app} K'
8	Compute the concentration of substrate at 50% saturation		E	$S_{0.5}$
	To compute best fit straight line for plotting:			
9	Enter substrate concentration to compute $\log [v/(V_{max} - v)]$	S	D R/S	$\log S$ $\log Y$

Program Listing--RPN

Line	Key	Comments	Line	Key	Comments
001	*LBLe		046	RCL4	
002	CLRG	Initialize: Clear	047	RCL6	
003	P⇄S	Registers	048	X	
004	CLRG		049	RCL9	
005	FIX		050	÷	
006	DSP3		051	-	
007	CLX		052	ENT↑	
008	RTN		053	ENT↑	
009	*LBLa		054	RCL4	
010	STO0	Store V_{max}	055	X²	
011	RTN		056	RCL9	
012	*LBLA		057	÷	
013	RCL0		058	RCL5	
014	X⇄Y		059	X⇄Y	
015	-		060	-	
016	LSTX		061	÷	
017	X⇄Y		062	STOB	
018	÷		063	X	
019	LOG	$S \to \log\ S$	064	RCL6	
020	X⇄Y		065	X²	
021	LOG	$v \to \log\ [v/(V_{max}- v)]$	066	RCL9	
022	F2?	Print if print flag	067	÷	
023	GSB9	is set	068	CHS	
024	Σ+	Accumulate sums	069	RCL7	Coefficient of
025	RTN		070	+	determination
026	*LBLB		071	÷	
027	RCL0	Deletion of data	072	PRTX	r^2
028	X⇄Y		073	RCL6	
029	-		074	RCL4	
030	LSTX		075	RCLB	
031	X⇄Y		076	X	
032	÷		077	-	
033	LOG		078	RCL9	
034	X⇄Y		079	÷	
035	LOG		080	STOA	
036	X⇄Y		081	RCLB	
037	F2?	Print Deletion	082	PRTX	Hill coefficient
038	GSB3	indicator if	083	X⇄Y	
039	F2?	print flag is set	084	CHS	
040	GSB9		085	10ˣ	
041	Σ-		086	SCI	
042	RTN		087	DSP3	
043	*LBLC	Compute regression	088	PRTX	Output K'
044	P⇄S	coefficients	089	SPC	
045	RCL8		090	P⇄S	

Line	Key	Comments
091	RTN	
092	*LBLD	Plotting:
093	FIX	$S \rightarrow \log S$
094	LOG	
095	R/S	
096	RCLB	
097	x	
098	RCLA	
099	+	
100	RTN	$\rightarrow \log [(v/V_{max} - v)]$
101	*LBLE	
102	RCLA	
103	CHS	
104	SCI	
105	10^x	
106	RCLB	
107	1/X	
108	Y^x	$\rightarrow S_{0.5}$
109	RTN	
110	*LBLd	Print flag: Set and
111	0	Clear
112	F2?	
113	RTN	
114	1	
115	SF2	
116	RTN	
117	*LBL9	Print subroutine
118	PRTX	
119	X⇄Y	
120	PRTX	
121	X⇄Y	
122	SF2	
123	SPC	
124	RTN	
125	*LBL3	Print deletion
126	SPC	indicator
127	DSP0	
128	1	
129	CHS	
130	PRTX	
131	SF2	
132	DSP3	
133	R↓	
134	RTN	

Register Contents, Labels, and Data Cards--RPN

Register	Contents	Labels	Contents
R_0	V_{max}	A	$S \uparrow v(+)$
R_{s4}	$\sum \log S$	B	$S \uparrow v(-)$
R_{s5}	$\sum (\log S)^2$	C	$\rightarrow r^2, n, K'$
R_{s6}	$\sum \log [v/(V_{max} - v)]$	D	$\log S \rightarrow \log (y)$
R_{s7}	$\sum (\log [v/(V_{max} - v)])^2$	E	$\rightarrow S_{0.5}$
R_{s8}	$\sum (\log S)(\log [v/(V_{max} - v)])$	a	$V_{max} \uparrow$
R_{s9}	n	d	Print?
R_A	$- \log K'$	e	Initialize
R_B	n_{app}		

User Instructions--Algebraic System

Step	Instructions	Input	Keys	Output
1	Print toggle		2nd B'	
2	Enter V_{max}	V_{max}	2nd A'	
3	Enter S	S	A	$\log S$
4	Enter into linear regression		R/S	
5	Enter v	v	B	$\log [v/V_{max} - v]$
6	Enter into linear regression enter all data pairs (S, v) before n_H calculation		R/S	n counter
7	Calculate n_{app}		C	n_{app}
8	Calculate correlation coefficient		D	r

Step	Instructions	Input	Keys	Output
9	Derive $\log\left(\dfrac{v}{V_{max}-v}\right)$	$\log S'$	E	$\log\left(\dfrac{v}{V_{max}-v}\right)$
10	Calculate $S_{0.5}$		R/S	$S_{0.5}$
11	Calculate K'		R/S	K'

PROGRAM LISTING--ALGEBRAIC SYSTEM

Line	Key	Entry	Comments	Line	Key	Entry	Comments
000	76	LBL		032	08	8	
001	17	B'	Print toggle	033	71	SBR	Call print
002	87	IFF		034	85	+	subroutine
003	08	08		035	43	RCL	
004	45	YX		036	29	29	
005	86	STF		037	28	LOG	
006	08	08		038	42	STO	
007	92	RTN		039	16	16	
008	76	LBL		040	42	STO	
009	45	YX		041	29	29	
010	22	INV		042	01	1	
011	86	STF		043	00	0	
012	08	08		044	71	SBR	Call print
013	92	RTN		045	85	+	subroutine
014	76	LBL	Enter V_{max}	046	43	RCL	
015	16	A'		047	16	16	
016	42	STO		048	91	R/S	
017	15	15		049	36	PGM	Enter linear
018	42	STO		050	01	01	regression
019	29	29		051	32	X:T	
020	00	0		052	92	RTN	
021	07	7		053	76	LBL	
022	71	SBR	Call print	054	12	B	Enter v
023	85	+	subroutine	055	42	STO	
024	43	RCL		056	17	17	
025	29	29		057	42	STO	
026	91	R/S		058	29	29	
027	76	LBL	Enter S	059	01	1	
028	11	A		060	01	1	
029	42	STO		061	71	SBR	Call print
030	29	29		062	85	+	subroutine
031	00	0		063	43	RCL	

Line	Key	Entry	Comments	Line	Key	Entry	Comments
064	17	17		107	14	D	
065	55	÷		108	69	OP	r
066	53	(		109	13	13	
067	43	RCL		110	42	STO	
068	15	15		111	29	29	
069	75	-		112	02	2	Call print subroutine
070	43	RCL		113	06	6	
071	17	17		114	71	SBR	
072	54	)	$\log\left(\dfrac{v}{V_{max}-v}\right)$	115	85	+	
073	95	=		116	43	RCL	
074	28	LOG		117	29	29	
075	42	STO		118	91	R/S	
076	29	29		119	76	LBL	$\log S$
077	01	1		120	15	E	$\rightarrow \log\left(\dfrac{v}{V_{max}-v}\right)$
078	02	2	Call print subroutine	121	69	OP	
079	71	SBR		122	14	14	
080	85	+		123	91	R/S	
081	01	1		124	00	0	
082	03	3	Call print subroutine	125	42	STO	
083	71	SBR		126	20	20	
084	85	+		127	43	RCL	
085	43	RCL		128	20	20	
086	29	29		129	69	OP	
087	91	R/S	Linear regression	130	15	15	
088	78	Σ+		131	42	STO	
089	92	RTN		132	21	21	
090	76	LBL		133	01	1	
091	13	C	n_{app}	134	00	0	
092	69	OP		135	45	Y$^{\times}$	
093	12	12		136	43	RCL	
094	32	X:T		137	21	21	$S_{0.5}$
095	42	STO		138	95	=	
096	19	19		139	42	STO	
097	42	STO		140	23	23	
098	29	29		141	42	STO	Call print subroutine
099	01	1	Call print subroutine	142	29	29	
100	04	4		143	02	2	
101	71	SBR		144	07	7	
102	85	+		145	71	SBR	
103	43	RCL		146	85	+	
104	19	19		147	43	RCL	
105	91	R/S		148	23	23	
106	76	LBL		149	91	R/S	

Line	Key	Entry	Comments
150	43	RCL	
151	23	23	
152	45	Y×	
153	43	RCL	
154	19	19	
155	95	=	K'
156	42	STO	
157	29	29	
158	02	2	
159	08	8	Call print
160	71	SBR	subroutine
161	85	+	
162	43	RCL	
163	29	29	
164	91	R/S	
165	76	LBL	
166	85	+	
167	87	IFF	
168	08	08	Print sub-
169	95	=	routine
170	43	RCL	
171	29	29	
172	92	RTN	
173	76	LBL	
174	95	=	
175	42	STO	
176	25	25	
177	73	RC*	
178	25	25	
179	69	OP	
180	04	04	
181	43	RCL	
182	29	29	
183	69	OP	
184	06	06	
185	69	OP	
186	00	00	
187	98	ADV	
188	92	RTN	

Register Contents, Labels, and Data Cards--Algebraic System

Register	Contents	Labels	Contents
R0 $\rightarrow$ R6	Linear regression	Label A'	V_{max}
R15	V_{max}	Label A	$S \rightarrow \log S$
R16	S	Label B	$v \rightarrow \log\left(v/V_{max} - v\right)$
R17	v	Label C	n_{app}
R19	n_{app}	Label	r
R20	$\log [v/V_{max} - v]$	Label E	$\log S' \rightarrow \log v/V_{max} - v$
R21	$\log S$		
R23	$S_{0.5}$		
R24	$1/n_{app}$		
R25	K'		

Data Card (4 2nd Write)

Alpha	Print Code	Register	Contents
42301344.	07	VMAX
36000000.	08	S
27322236.	10	LOGS
42000000.	11	V
27322242.	12	LOGV
42302042.	13	VM-V
31230000.	14	NH
15323535.	26	CORR
36014006.	27	SO.5
26650000.	28	K'

Example

The following data were obtained for an enzyme-catalyzed reaction exhibiting sigmoidal kinetics with V_{max} = 100 μmoles/liter·min.   Calculate K', n (n_{app}), and $S_{0.5}$.

S (M)	V (μmolers/liter·min)
6.25×10^{-4}	1.54
12.50×10^{-4}	5.88
25×10^{-4}	20
50×10^{-4}	50
100×10^{-4}	80
200×10^{-4}	94.12
400×10^{-4}	98.46
800×10^{-4}	99.61

Solution:

$$r^2 = 1.000$$
$$n\ (n_{app}) = 2.00$$
$$K' = 2.5 \times 10^{-5}$$
$$S_{0.5} = 5 \times 10^{-3}$$

References

1. I.H. Segel (1976), *Biochemical Caluclations*, 2nd Ed., John Wiley & Sons, Inc., New York, pp. 309-311.

2. I.H. Segel (1976), *Enzyme Kinetics*, John Wiley & Sons, Inc., New York, Chapter 7.

3. A. Cornish-Bowden and D.E. Koshland (1975), "Diagnostic Uses of the Hill (Logit and Nernst) Plots," *Journal of Molecular Biology 95*, 201-212.

4. A.V. Hill (1910), *Journal of Physiology 40*, London, 4.

5. G.R. Ainslie, Jr., J.P. Shill, and K.E. Neet (1972), *Journal of Biological Chemistry 247*, 7088-7096.

4C. FIRST AND SECOND ORDER CHEMICAL KINETICS

First Order Kinetics

First order reactions are quite common to biochemistry. Some examples are to be found in radioactive tracer studies and single turnover type enzyme kinetics. The simplest reaction involves a reactant A being converted to a product P.

$$A \overset{k}{\rightarrow} P \tag{1}$$

Since the reaction has a single component, the rate equation takes the form of the velocity equaling the rate constant times the concentration of the reactant:

$$\frac{d\mathrm{P}}{dt} = K[\mathrm{A}] \tag{2}$$

Equation 2 can be rearranged:

$$-\frac{d[\mathrm{A}]}{[\mathrm{A}]} = k\ dt \tag{3}$$

and integrated within definite limits:

$$\int_{\mathrm{A-X}}^{\mathrm{A}} \frac{d[\mathrm{A}]}{[\mathrm{A}]} = k \int_{0}^{t} dt \tag{4}$$

$$\ln \mathrm{A} - \ln (\mathrm{A} - \mathrm{X}) = k(t - 0) \tag{5}$$

to obtain

$$\ln \frac{\mathrm{A}}{\mathrm{A-X}} = kt \tag{6}$$

In equation 6 the expression A - X is the concentration of A at time t (i.e., the initial amount of reactant A minus X, the amount of A which has been transformed to product).

The calculator program allows transformation of the initial and time course concentrations of reactant A into the ln (A/(A - X)) format and enters the data into the linear regression subroutine. The rate constant can then be determined from the slope of the fitted line. The error in the constant can be estimated from the correlation coefficient for the fitted line, as well as from graphical inspection of experimental values compared with derived estimates for plots of ln $(A/(A - X))$ vs. t (Figure 4C1). Frequently the half time $(t_{\frac{1}{2}})$ for a first order reaction is of interest. The half-time or half-life of a reaction is the amount of time for the reactant (A) to halve its initial concentration $(X = A/2)$. Half-time can be determined by substituting $X = A/2$ in equation 6 and solving for t:

$$t = \frac{1}{k} \ln \frac{A}{A/2} = \frac{\ln 2}{k} = \frac{0.693}{k}$$

When a reaction end point is not obvious because of instrumental drift or because of other, slower reactions, it is desirable to estimate the first order rate constant from a single progress curve without determining the asymptote. This can be accomplished using the Guggenheim algebraic procedure, which takes the natural log of the difference between early observations $(A_1, A_2, \ldots)$ and values approximately three half-times from zero time $(A_1', A_2', \ldots)$ (Figure 4C2). The first order rate constant can then be determined from the slope of the line of a plot of ln $(A_i - A_i')$ vs. t. Apparent first order behavior is one thing, while actual kinetic mechanism is quite another. This is particularly true when in a biochemical reaction one component is in excess $(A + B \rightarrow P$, where B is in excess), thereby giving a mechanistically second order reaction pseudo first order kinetics with velocities dependent on the concentration of the minor component. Another instance of pseudo first order kinetics involves bimolecular cyclic reactions where one reactant is generated faster than it is utilized, thus effectively maintaining a constant concentration for that component in the reaction.

Regardless of whether reaction kinetics are first order or pseudo first order, kinetic constants, reaction order, and half-time still provide important guide posts in the characterization of biochemical reactions.

User Instructions--RPN

Step	Instructions	Input	Keys	Output
1	Initialize: Clear Registers		f e	0.00
2	To set print flag		f d	1.00
3	To clear print flag		f d	0.00
	For standard first order kinetics:			
4	Enter infinity value of A	A_∞	f a	A_∞
5	Enter: time	t	ENTER↑	t
	absorbance	X	A	n
6	Repeat for all data			
7	Compute coefficient of determination and rate constant		C	r^2 k
8	Compute half-time of reaction		D	$t_{\frac{1}{2}}$
	For Guggenheim method:			
9	Enter: time	t	ENTER↑	t
	A_1	A_1	ENTER↑	A_1
	A_1'	A_1'	B	n
10	Compute coefficient of determination and rate constant		C	r^2 k
11	Compute half-time of reaction		D	$t_{\frac{1}{2}}$
	To compute best fit straight line:			
12	Enter t		E	$\ln y$

Program Listing--RPN

Line	Key	Comments	Line	Key	Comments
001	*LBLe		046	RCL9	
002	CLRG	Initialize: Clear	047	÷	
003	P≷S	Registers	048	RCL5	
004	CLRG		049	X≷Y	
005	CLX		050	-	
006	DSP3		051	÷	
007	RTN		052	STOB	
008	*LBLa		053	x	
009	STOA	Store A_∞	054	RCL6	
010	RTN		055	X²	
011	*LBLA	Standard first	056	RCL9	
012	RCLA	order kinetics	057	÷	
013	X≷Y	Enter: t	058	CHS	
014	-	X	059	RCL7	
015	RCLA		060	+	
016	X≷Y		061	÷	
017	÷		062	STOC	
018	LN	→ $\ln[A_\infty/(A_\infty - X)]$	063	PRTX	→ r^2
019	X≷Y		064	RCL6	
020	F2?	Print if print flag	065	RCL4	
021	GSB9	is on	066	RCLB	
022	Σ+	Accumulate sums	067	x	
023	RTN		068	-	
024	*LBLB	Guggenheim Method:	069	RCL9	
025	-	Enter: t	070	÷	
026	ABS	A_i	071	STOA	
027	LN	A'_i	072	RCLB	
028	X≷Y	$\ln(A'_i - A_i)$	073	ABS	
029	F2?		074	PRTX	
030	GSB9	Print if print flag	075	P≷S	
031	Σ+	is on	076	RTN	→ k
032	RTN	Accumulate sums	077	*LBLD	
033	*LBLC		078	2	
034	P≷S		079	LN	
035	RCL8	Compute coefficient	080	RCLB	Compute half-time
036	RCL4	of determination	081	ABS	for reaction
037	RCL6	and rate constant	082	÷	
038	x		083	DSP0	
039	RCL9		084	PRTX	→ $t_{\frac{1}{2}}$
040	÷		085	SPC	
041	-		086	RTN	
042	ENT↑		087	*LBLE	Plotting/projection
043	ENT↑		088	DSP3	routine
044	RCL4		089	RCLB	
045	X²		090	x	

Line	Key	Comments
091	RCLA	
092	+	
093	RTN	
094	*LBLd	Print flag:
095	0	Set and Clear
096	F2?	
097	RTN	
098	1	
099	SF2	
100	RTN	
101	*LBL9	Print subroutine
102	PRTX	
103	X⇄Y	
104	PRTX	
105	X⇄Y	
106	SF2	
107	SPC	
108	RTN	

Register Contents, Labels, and Data Cards--RPN

Register	Contents	Labels	Contents
R_{s4}	$\sum t$	A	$t \uparrow X(+)$
R_{s5}	$\sum t^2$	B	$T \uparrow A_1 \uparrow A_1'(+)$
R_{s6}	$\sum \ln y$	C	$\rightarrow r^2, k$
R_{s7}	$\sum (\ln y)^2$	D	$\rightarrow t_{\frac{1}{2}}$
R_{s8}	$\sum \ln y\ (t)$	E	$t \rightarrow \ln y$
R_{s9}	n	a	$A_\infty \uparrow$
R_A	A_∞	d	Print?
R_B	k	e	Initialize
R_C	r^2		

User Instruction--Algebraic Systems

Step	Instructions	Input	Keys	Output
1	Print toggle		2nd B'	
2	Enter infinity value	A	2nd A'	
3	Enter time	t	A	
4	Enter concentration, X	S	B	$\ln[A/(A - X)]$
5	Enter data into linear regression Enter all data pairs before calculation of k		R/S	n

Step	Instructions	Input	Keys	Output
6	Calculate rate constant		C	k
7	Determine correlation coefficient		D	r
8	To derive fitted line enter	t'	E	$\ln[A/(A - X)]'$
9	Determine $t_{1/2}$		R/S	$t_{1/2}$

Program Listing--Algebraic System

Line	Key	Entry	Comments	Line	Key	Entry	Comments
000	76	LBL		029	42	STO	
001	17	B'	Print	030	29	29	Call print
002	87	IFF	toggle	031	02	2	subroutine
003	08	08		032	02	2	
004	45	YX		033	71	SBR	
005	86	STF		034	85	+	
006	08	08		035	43	RCL	
007	92	RTN		036	29	29	
008	76	LBL		037	36	PGM	Linear
009	45	YX		038	01	01	regression
010	22	INV		039	32	X:T	
011	86	STF		040	92	RTN	Enter X
012	08	08		041	76	LBL	
013	92	RTN		042	12	B	
014	76	LBL	Enter A	043	42	STO	
015	16	A'		044	19	19	Call print
016	42	STO		045	42	STO	subroutine
017	16	16	Call print	046	29	29	
018	42	STO	subroutine	047	02	2	
019	29	29		048	03	3	
020	02	2		049	71	SBR	Convert to
021	01	1		050	85	+	$\ln (A/A - X)$
022	71	SBR		051	43	RCL	
023	85	+	Enter t	052	16	16	
024	91	R/S		053	55	÷	
025	76	LBL		054	53	(	
026	11	A		055	43	RCL	
027	42	STO		056	16	16	
028	17	17		057	75	-	

Line	Key	Entry	Comments	Line	Key	Entry	Comments
058	43	RCL		100	43	RCL	
059	19	19		101	29	29	
060	54	)		102	91	R/S	
061	95	=		103	76	LBL	Enter t'
062	23	LNX		104	15	E	
063	42	STO		105	42	STO	
064	29	29		106	21	21	
065	02	2		107	42	STO	Derive
066	04	4		108	29	29	ln $A/A - X'$
067	71	SBR	Call print	109	02	2	
068	85	+	subroutine	110	07	7	
069	43	RCL		111	71	SBR	Call print
070	29	29		112	85	+	subroutine
071	91	R/S		113	43	RCL	
072	78	Σ+	Linear	114	29	29	
073	91	R/S	regression	115	69	OP	
074	76	LBL		116	14	14	
075	13	C	Calculate k	117	42	STO	
076	69	OP		118	29	29	
077	12	12		119	02	2	
078	32	X:T		120	08	8	
079	42	STO		121	71	SBR	Call print
080	20	20		122	85	+	subroutine
081	42	STO		123	43	RCL	
082	29	29		124	29	29	
083	00	0		125	91	R/S	
084	08	8		126	93	.	
085	71	SBR	Call print	127	06	6	Determine $t_{\frac{1}{2}}$
086	85	+	subroutine	128	09	9	
087	43	RCL		129	03	3	
088	29	29		130	55	÷	
089	91	R/S		131	43	RCL	
090	76	LBL		132	20	20	
091	14	D	r	133	95	=	
092	69	OP		134	42	STO	
093	13	13		135	29	29	
094	42	STO		136	00	0	
095	29	29		137	07	7	
096	02	2		138	71	SBR	Call print
097	06	6		139	85	+	subroutine
098	71	SBR	Call print	140	91	R/S	
099	85	+	subroutine	141	76	LBL	

Line	Key	Entry	Comments
142	85	+	Print
143	87	IFF	subroutine
144	08	08	
145	95	=	
146	43	RCL	
147	29	29	
148	92	RTN	
149	76	LBL	
150	95	=	
151	42	STO	
152	25	25	
153	73	RC*	
154	25	25	
155	69	OP	
156	04	04	
157	43	RCL	
158	29	29	
159	69	OP	
160	06	06	
161	69	OP	
162	00	00	
163	98	ADV	
164	92	RTN	

Register Contents, Labels, and Data Cards--Algebraic System

Register	Contents	Labels	Contents
R0 → R6	Linear regression	Label A'	Enter A
R16	A	Label B'	Print toggle
R17	t	Label A	Enter t
R10	X	Label B	Enter X
R20	k	Label C	Calculate k
R21	t'	Label D	r
		Label E	Enter t' Derive ln $[A/(A - X')]$

Data Card (4 2nd Write)

Alpha Print Code	Register	Contents
37026303.	07	T1/2
26000000.	08	K
13000000.	21	A
37243017.	22	TIME
44000000.	23	X
27311363.	24	LNA/
15323535.	26	CORR
37650000.	27	T*
27311365.	28	LNA*

User Instructions--Algebraic System (Guggenheim)

Step	Instructions	Input	Keys	Output
1	Enter time	t	A	
2	Enter concentration	A	B	
3	Enter concentration	A'	R/S	$\ln(A - A')$
4	Enter data into linear regression		R/S	n
	Enter all data pairs before calculation of rate constant			
5	Calculate rate constant		C	k
6	Determine correlation coefficient		D	r
7	To derive fitted line enter t'	t'	E	$\ln(A - A')$
8	Determine $t_{\frac{1}{2}}$		R/S	$t_{\frac{1}{2}}$

Program Listing--Algebraic System

Line	Key	Entry	Comments	Line	Key	Entry	Comments
000	76	LBL		041	42	STO	
001	11	A	Enter t	042	20	20	
002	42	STO		043	91	R/S	
003	16	16		044	76	LBL	
004	76	LBL		045	14	D	
005	32	X:T		046	69	OP	Enter t'
006	36	PGM		047	13	13	
007	01	01		048	91	R/S	Derive
008	32	X:T		049	76	LBL	$\ln (A - A')$
009	92	RTN		050	15	E	
010	91	R/S		051	42	STO	Determine $t_{\frac{1}{2}}$
011	76	LBL		052	21	21	
012	12	B		053	69	OP	
013	42	STO	Enter A_1	054	14	14	
014	17	17		055	91	R/S	
015	91	R/S		056	93	.	
016	42	STO		057	06	6	
017	18	18	Enter A_1'	058	09	9	
018	43	RCL		059	03	3	
019	17	17	Convert to	060	42	STO	
020	75	-	$\ln (A_1 - A_1')$	061	22	22	
021	43	RCL		062	55	÷	
022	18	18		063	43	RCL	
023	95	=		064	20	20	
024	77	GE		065	95	=	
025	39	COS		066	91	R/S	
026	94	+/-					
027	76	LBL					
028	39	COS	Call linear				
029	23	LNX	regression				
030	42	STO					
031	19	19					
032	91	R/S					
033	78	Σ+	Calculate k				
034	92	RTN					
035	91	R/S					
036	76	LBL					
037	13	C					
038	69	OP	Calculate r				
039	12	12					
040	32	X:T					

Register Contents, Labels, and Data Cards--Algebraic System

Register	Contents	Labels	Contents
R0 → R6	Linear regression	Label A	Enter t
R17	A_1	Label B	Enter A_1
R20	k	Label C	Calculate k
R21	t'	Label D	Calculate r
		Label E	Enter t'; derive $\ln (A - A')$

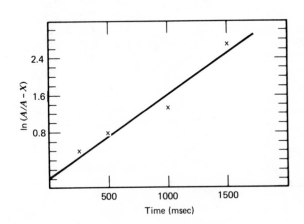

Figure 4C1. First order kinetics--plot of $\ln (A/A - X)$ versus t.

Example

The reaction of carbon monoxide with cytochrome oxidase was followed at -75°C. The CO was initially bound to mitochondrial cytochrome oxidase, and the reaction was initiated by photodissociation. The recombination kinetics was then monitored spectroscopically at 445 nm. Calculate the first order rate constant and the half-time for the reaction.

TABLE 1

$$A_\infty = 13.2\% \, T^*$$

t(msec)	$X(\Delta\%T)$
250	3.1
500	6.2
1000	9.3
1500	12.2

Solution

t(msec)	$\ln [A/(A - X)]$
250	0.268
500	0.634
1000	1.219
1500	2.580

$k = 0.002$ msec^{-1}

$t_{\frac{1}{2}} = 386.18$ msec

$*\%T$ = percent transmittance

Fitted line

t'	$\ln A/A - X$
500	0.615
1500	2.409

$r = 0.979$

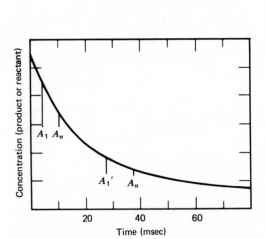

Figure 4C2.   Sampling procedure for the Guggenheim method.

Example

The kinetics of mitochondrial cytochrome c oxidation can be described using first order formalism. In this example the substrate-reduced, anaerobic mitochondria are mixed with a pulse of oxygenated buffer, and the oxidation of cytochrome c is followed spectroscopically at 550 nm.   Calculate the rate constant using the Guggenheim method.

TABLE 1

t(msec)	$A(\Delta\%T)$	$A'(\Delta\%T)$
0	0	10.24
25	1.28	10.56
50	3.20	10.88
75	4.80	12.16
100	6.40	12.48

Solution

t(msec)	$\ln(A - A')$
0	2.326
25	2.228
50	2.039
75	1.996
100	1.805

$k = 0.005 \ \text{msec}^{-1}$
$r = 0.986$

t (msec)	$\ln (A - A')$
50	2.079
100	1.824

$t_{\frac{1}{2}}$ = 136 msec

Second Order Kinetics

When a reaction rate is influenced by two reactants, the reaction is termed second order. Bimolecular or second order reactions can involve the reaction of two similar molecules $(A + A \rightarrow P)$ or the reaction of two differing species $(A + B \rightarrow P)$. To obtain the rate constant for a reaction of two differing molecules, the rate equation (equation 1), where x is the amount of the reactants used for time t, and a and b are initial concentrations of the reactants:

$$\frac{dx}{dt} = k(a - x)(b - x) \tag{1}$$

is arranged as follows:

$$\frac{dx}{(a - x)(b - x)} = k \, dt \tag{2}$$

Since $1/(a - x)(b - x)$ is algebraically equivalent to $[(1/(a - b)][1/(b - x) - 1/(a - x)]$, the latter can be substituted into equation 2 to give

$$\frac{1}{a - b} \int_0^x \left(\frac{dx}{b - x} - \frac{dx}{a - x} \right) = k \int_0^t dt \tag{3}$$

and expression 3 can be integrated using the $(a + bx)$ integral
$\int dx/(a+bx) = 1/b \, [\ln (a + bx)] + c$:

$$\frac{1}{a - b} \ln \left(\frac{a - x}{b - x} \right) - \frac{1}{a - b} \left(\ln \frac{a}{b} \right) = kt \tag{4}$$

$$\frac{1}{a - b} \ln \left[\frac{b(a - x)}{a(b - x)} \right] = kt \tag{5}$$

Equation 5 is that of a straight line and can be evaluated statistically and graphically by using a data transform-linear regression program. The rate constant can be determined by plotting $\ln [b(a - x)/a(b - x)]$ versus time to obtain the slope and then by dividing by $a - b$.

A special case of second order reactions is encountered when one of the reactants is present in excess. If component B is in excess, the second order equation will be reduced to $dx/dt = kb(a - x)$ and the reaction will follow apparent first order kinetics. Another example of second order kinetics occurs when the reaction is between two similar molecules ($A + A \rightarrow P$) or the initial concentrations of two differing molecules are the same ($A = B$, where $A + B \rightarrow P_a + P_b$). In this case the differential equation takes the following form

$$\frac{dx}{dt} = k(a - x)^2 \tag{6}$$

The integrated rate (equation 7) is a straight line plot with a slope equal to the rate constant:

$$\frac{x}{a(a - x)^2} = kt \tag{7}$$

The half-time for the reaction is the reciprocal of the product of the rate constant and the initial concentration:

$$t_{\frac{1}{2}} = \frac{1}{k_2 a} = \frac{\frac{1}{2}a}{k_a(\frac{1}{2}a)} \tag{8}$$

User Instructions--RPN

Step	Instructions	Input	Keys	Output
1	Initialize: Clear Registers		f e	0.00
2	To set print flag		f d	1.00
3	To clear print flag		f d	0.00
4	Enter initial concentration of reactant	A	ENTER↑	
5	Enter initial concentration of reactant B	B	f a	
6	Enter time t Enter concentration of product at time t	t X	ENTER↑ A	n
7	To delete data, enter: t X	t X	ENTER↑ B	$n - 1$
8	Compute coefficient of determination and second order rate constant, k_2		C	r^2 k_2
9	Compute half-time for reaction, $t_{\frac{1}{2}}$		D	$t_{\frac{1}{2}}$
10	For projection of best fit line, enter t	t	E	$\ln y$

Program Listing--RPN

Line	Key	Comments	Line	Key	Comments
001	*LBLe		046	SPC	
002	CLRG	Initialize: Clear	047	RCL8	Compute coefficient
003	P⇄S	Registers	048	RCL4	of determination and
004	CLRG		049	RCL6	k_2
005	CLX		050	x	
006	CF0		051	RCL9	
007	CF1		052	÷	
008	CF2		053	-	
009	RTN		054	ENT↑	
010	*LBLa		055	ENT↑	
011	STO1	Store B	056	RCL4	
012	X=Y?	$A = B$?	057	X²	
013	SF0	Yes: set flag	058	RCL9	
014	R↓	No: continue	059	÷	
015	STO0	Store A	060	RCL5	
016	RTN		061	X⇄Y	
017	*LBLA		062	-	
018	STOD	Store x	063	÷	
019	R↓		064	STOB	→ slope
020	STOE	Store t	065	x	
021	F0?		066	RCL6	
022	GTO0		067	X²	
023	GTO1		068	RCL9	
024	*LBL7		069	÷	
025	F1?		070	CHS	
026	GSB9		071	RCL7	
027	Σ+	Accumulate sums	072	+	
028	RTN		073	÷	
029	*LBLB		074	PRTX	→ r^2
030	STOD	Delete data routine	075	RCL6	
031	R↓		076	RCL4	
032	STOE		077	RCLB	
033	SF2		078	x	
034	F0?		079	-	
035	GTO0		080	RCL9	
036	GTO1		081	÷	
037	*LBL8		082	STOA	→ intercept
038	F1?		083	P⇄S	
039	GSB3		084	RCLB	
040	F1?		085	F0?	
041	GSB9		086	GTO2	
042	Σ-	Data deletion	087	RCL0	
043	RTN		088	RCL1	
044	*LBLC		089	-	
045	P⇄S		090	÷	

Line	Key	Comments	Line	Key	Comments
091	PRTX	Output rate constant	138	÷	
092	RTN		139	LN	
093	*LBL2		140	RCLB	
094	SCI		141	÷	
095	PRTX		142	FIX	
096	RTN		143	DSP2	
097	*LBLD	Calculate half-time	144	PRTX	→ $t_{\frac{1}{2}}$
098	F0?	for reaction	145	RTN	
099	GT04		146	*LBL3	Print deletion indi-
100	RCL1		147	SPC	cator
101	RCL0		148	DSP0	
102	X>Y?		149	1	
103	GT05		150	CHS	
104	X⇄Y		151	PRTX	
105	2	Half-time calculation	152	DSP2	
106	×	for $A < B$	153	R↓	
107	X⇄Y		154	RTN	
108	-		155	*LBL9	Print subroutine
109	RCL1		156	PRTX	
110	X⇄Y		157	X⇄Y	
111	÷		158	PRTX	
112	LN		159	X⇄Y	
113	RCL0		160	SPC	
114	RCL1		161	RTN	
115	-		162	*LBLE	Projection routine
116	RCLB		163	DSP3	for $A \neq B$
117	×		164	F0?	
118	÷		165	GT06	
119	FIX		166	RCLB	
120	DSP2		167	×	
121	PRTX	→ $t_{\frac{1}{2}}$	168	RCLA	
122	RTN		169	+	
123	*LBL4		170	RTN	
124	RCLB	Half-time calculation	171	*LBL6	Projection subrou-
125	RCL0	for $A = B$	172	RCLB	tine for $A = B$
126	×		173	×	
127	1/X		174	RCLA	
128	FIX		175	+	
129	DSP2		176	RTN	
130	PRTX	→ $t_{\frac{1}{2}}$	177	*LBLd	Print flag:
131	RTN		178	F1?	Set and Clear
132	*LBL5	Half-time calculation	179	GT02	
133	2	for $A > B$	180	SF1	
134	×		181	1	
135	X⇄Y		182	RTN	
136	-		183	*LBL2	
137	RCL0		184	0	

Line	Key	Comments
185	CF1	
186	RTN	
187	*LBL0	Subroutine for
188	RCLD	$A = B$
189	ENT↑	
190	ENT↑	
191	RCL0	
192	X⇄Y	
193	-	
194	RCL0	
195	x	
196	÷	
197	RCLE	
198	F2?	
199	GTO8	
200	GTO7	
201	*LBL1	Subroutine for
202	RCL0	$A \neq B$
203	RCLD	
204	-	
205	RCL1	
206	x	
207	LSTX	
208	RCLD	
209	-	
210	RCL0	
211	x	
212	÷	
213	LN	
214	RCLE	
215	F2?	
216	GTO8	
217	GTO7	

Register Contents, Labels, and Data Cards--RPN

Register	Contents		Labels	Contents
R_0	A		A	$t \uparrow x\ (+)$
R_1	B		B	$t \uparrow x\ (-)$
R_{s4}	$\sum x$		C	$\rightarrow r^2,\ k_2$
R_{s5}	$\sum x^2$		D	$\rightarrow t_{\frac{1}{2}}$
R_{s6}	$\sum y$		E	$t' \rightarrow \ln y$
R_{s7}	$\sum y^2$		a	$A \uparrow B$
R_{s8}	$\sum xy$		d	Print?
R_{s9}	n		e	Initialize
R_A	Intercept			
R_B	k_2			
R_D	Last x			
R_E	Last t			

User Instructions--Algebraic System

Step	Instructions	Input	Keys	Output
1	Enter initial concentration of reactant A	a	2nd A'	
2	Enter initial concentration of reactant B	b	R/S	
3	Enter time	t	A	
4	Enter concentration, x	x	B	$\ln[b(a - x)/a(b - x)]$

Step	Instructions	Input	Keys	Output
5	Enter data into linear regression subroutine		R/S	n
	Enter all data pairs before calculation of rate constant			
6	Calculate rate constant		C	k
7	Determine correlation coefficient		D	r
8	To derive fitted line enter t'	t'	E	ln $[b(a - x)/a(b - x)]$

Program Listing--Algebraic System

Line	Key	Entry	Comments	Line	Key	Entry	Comments
000	76	LBL		024	53	(	
001	16	A'		025	43	RCL	Convert to
002	42	STO	Enter a	026	11	11	$\ln\left(\dfrac{b(a - x)}{a(b - x)}\right)$
003	10	10		027	65	×	
004	91	R/S		028	53	(	
005	42	STO	Enter b	029	43	RCL	
006	11	11		030	10	10	
007	91	R/S		031	75	-	
008	76	LBL		032	43	RCL	
009	11	A		033	14	14	
010	42	STO	Enter t	034	54	)	
011	12	12		035	55	÷	
012	76	LBL		036	53	(	
013	32	X:T		037	53	(	
014	36	PGM		038	43	RCL	
015	01	01		039	10	10	
016	32	X:T		040	65	×	
017	92	RTN		041	53	(	
018	91	R/S		042	43	RCL	
019	76	LBL		043	11	11	
020	12	B	Enter x	044	75	-	
021	42	STO		045	43	RCL	
022	14	14		046	14	14	
023	53	(		047	54	)	

Line	Key	Entry	Comments
048	54	)	
049	95	=	
050	23	LNX	
051	91	R/S	
052	78	Σ+	Call linear
053	92	RTN	regression
054	91	R/S	
055	76	LBL	
056	13	C	Calculate k_2
057	69	OP	
058	12	12	
059	32	X:T	
060	42	STO	
061	20	20	
062	55	÷	
063	53	(	
064	43	RCL	
065	10	10	
066	75	-	
067	43	RCL	
068	11	11	
069	54	)	
070	95	=	
071	42	STO	
072	21	21	
073	91	R/S	
074	76	LBL	
075	14	D	Calculate r
076	69	OP	
077	13	13	
078	91	R/S	
079	76	LBL	
080	15	E	Enter t'
081	69	OP	Calculate
082	14	14	
083	91	R/S	$\ln\left(\dfrac{b(a - x)}{a(b - x)}\right)$

Register Contents, Labels and Data Cards--Algebraic System

Register	Contents	Labels	Contents
RO → R6	Linear regression	Label A'	Enter a
R10	a	Label A	Enter t
R11	b	Label B	Enter x
R14	x	Label C	Calculate k
		Label D	Calculate r
		Label E	Enter t'; Calculate ln $b(a - x)/a(b - x)$

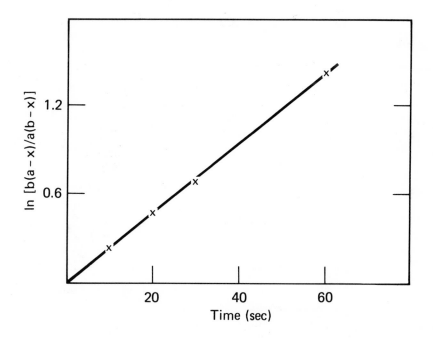

Figure 4C3.   Second order plot where $A = B$.

Example

From the data given in Table 1 determine the second order
rate constant for the reaction between compounds A and B.
Initial concentration of reaction A = 0.2 M, B = 0.1 M

TABLE 1

Time (sec)	x amount converted to product
10	0.034
20	0.054
30	0.066
60	0.086

Solution

t (sec)	$\ln [b(a - x)/a(b - x)]$
10	0.229
20	0.462
30	0.678
60	1.404

$r = 0.9998$
$k = 0.235$ liter/mole·sec

Fitted line

t'	$\ln [b(a - x)/a(b - x)]$
10	0.223
60	1.399

User Instructions--Algebraic System

Step	Instructions	Input	Keys	Output
1	Enter infinity value	a	2nd A'	
2	Enter time	t	A	
3	Enter concentration, x	x	B	$x/a(a - x)$
4	Enter data into linear regression		R/S	n
	Enter all data pairs before calculation of rate constant			
5	Calculate rate constant		C	k
6	Determine correlation coefficient		D	r
7	To derive fitted line enter t'	t'	E	$x/a(a - x)$
8	Determine $t_{\frac{1}{2}}$		R/S	$t_{\frac{1}{2}}$

Program Listing--Algebraic System

Line	Key	Entry	Comments	Line	Key	Entry	Comments
000	76	LBL		041	13	C	
001	16	A'	Enter a	042	69	OP	
002	42	STO		043	12	12	
003	16	16		044	32	X:T	
004	91	R/S		045	42	STO	
005	76	LBL		046	20	20	Calculate k
006	11	A	Enter t	047	91	R/S	
007	42	STO		048	76	LBL	
008	17	17		049	14	D	
009	76	LBL		050	69	OP	
010	32	X:T		051	13	13	
011	36	PGM		052	91	R/S	Calculate r
012	01	01		053	76	LBL	
013	32	X:T		054	15	E	
014	92	RTN		055	69	OP	
015	91	R/S		056	14	14	
016	76	LBL		057	91	R/S	
017	12	B	Enter x	058	43	RCL	
018	42	STO		059	16	16	Enter t'
019	18	18	Convert to	060	65	×	
020	55	÷	ln $x/a(a-x)$	061	43	RCL	Calculate
021	53	(		062	20	20	ln $x/a(a-x)$'
022	43	RCL		063	95	=	
023	16	16		064	35	1/X	Determine $t_{\frac{1}{2}}$
024	65	×		065	42	STO	
025	53	(		066	21	21	
026	43	RCL		067	91	R/S	
027	16	16					
028	75	-					
029	43	RCL					
030	18	18					
031	54	)					
032	54	)					
033	95	=					
034	42	STO					
035	19	19					
036	91	R/S					
037	78	Σ+	Call linear				
038	92	RTN	regression				
039	91	R/S					
040	76	LBL					

Register Contents, Labels, and Data Cards--Algebraic System

Register	Contents	Labels	Contents
R0 → R6	Linear regression	Label A'	Enter a
R16	a	Label A	Enter t
R18	x	Label B	Enter x
R20	k	Label C	Calculate k
		Label D	Calculate r
		Label E	Enter t'; calculate ln $x/a(a - x)$

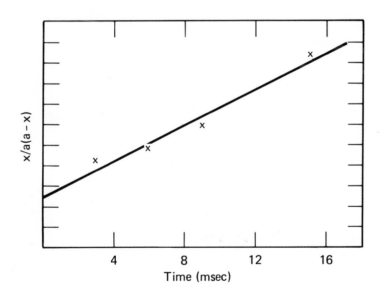

Figure 4C4. Second order plot where $A \neq B$.

Example

The kinetics of cytochrome c_2 oxidation in *Rhodopseudomonas sphaeroides Ga* chromatophores was followed at 550 nm after a flash of actinic light. Assuming that the electron donor-acceptor pair concentrations are equal (i.e., c_2^+, ZH_2), the reaction can be described using a second order plot (equation 7).

TABLE 1

Time (msec)	ΔA^*, 550–540 nm
3	0.00033
6	0.00067
9	0.00083
15	0.00093

$\Delta A_\infty = 0.001$

Solution

Time (msec)	$x/a(a - x)$
3	492.5
6	2,030.0
9	4,882.4
15	13,285.7

$k_2 = 1.094 \times 10^3$ mole^{-1} sec^{-1}

$r = 0.983$

$t_{\frac{1}{2}} = 0.91$ msec

*ΔA = change in absorbance

Fitted Line

t' (msec)	$x/a\,a - x)$
4	521.4
15	12560.1

References

1. W. Moore (1962), *Physical Chemistry*, Longmans-Prentice-Hall, Inc., New York, pp. 253-263.

2. H. Gutfreund (1972), *Enzymes: Physical Principles*, John Wiley & Sons, Inc., New York, pp. 116-120.

3. M. Erecinska and B. Chance (1972), *Archives of Biochemistry and Biophysics 151*, 304-315.

4. R. Prince, L. Bashford, K. Takamiya, W. van den Berg, and P. Dutton (1978), *Journal of Biological Chemistry 253*, 4137-4141.

4D. MICHAELIS-MENTEN ENZYME KINETICS--WEIGHTED LINEAR REGRESSION

In this section, we are mainly concerned with the methods used to analyze enzyme kinetic data to determine the Michaelis constants, K_m and V_{max}. We do not derive the Michaelis-Menten equation or any other rate equations since the theoretical basis of enzyme kinetics is well described in several textbooks.

For many years, biochemists have evaluated Michaelis constants, maximum velocities, and inhibition constants for enzyme catalyzed reactions by graphical analysis. If the concentrations of all reactants but one are held constant, for instance, the initial velocity of an enzymatic reaction often follows the rate law derived by Michaelis and Menten:

$$v = \frac{V_{max}S}{K_m + S}$$ (1)

where v = velocity
S = substrate concentration
K_m = the Michaelis constant
V_{max} = maximum velocity

Equation 1 represents a rectangular hyperbola passing through the origin with the line $V = V_{max}$ as a horizontal asymptote. Since this is not a convenient form for graphical analysis, the equation is usually rearranged into a linear form such as the familiar Lineweaver-Burk equation:

$$\frac{1}{v} = \frac{K_m}{V_{max}} + \frac{1}{V_{max}}$$ (2)

from which the parameters K_m and V_{max} are easily obtained from the slope and intercept of the resulting straight line.

Accurate and unambiguous evaluation of the Michaelis constants can be a demanding and tedious experimental task. The initial velocities of the enzymatic reaction at a series of substrate concentrations must be determined. This means that the substrate or product concentrations have to be measured as a function of time for each initial substrate concentration. These results are plotted as functions of time, and the slopes of the lines at zero time are determined; these are the initial reaction rates and are the rates to be used to evaluate the Michaelis constants. The graphical analysis involves plotting $1/v$ versus $1/S$ (equation 2) to determine the slope, K_m/V_{max}, and the intercept, $1/V_{max}$. Students have labored long and hard in laboratory

exercises to perform these graphical analyses by drawing the best
fit straight line through the points as determined by eye. However, the availability of pocket calculators at relatively low
cost, some with linear regression built in, have made it possible
to quickly and effortlessly determine the best fit straight line
based on statistical considerations and to obtain an estimate of
the reliability of the fitted constants.

Such an analysis is generally quite adequate, especially for
first approximations. However, among practicing enzyme kineti-
cists statistical weighting of the data is considered essential
if any meaningful conclusions are to be drawn from the computed
parameters.

Wilkinson (2), in analyzing the data of Atkinson et al. (4)
on the kinetics of the enzyme nicotinamide mononucleotide adenylyltransferase, determined the weighting factor to be the velocity to the fourth power (v^4) which is the reciprocal of the vari-
ance.

For the details of this derivation see Wilkinson's paper.
The equations used by Wilkinson to obtain provisional estimates
of K_m and V_{max} are as follows:

$$\alpha = \sum v^3 \qquad\qquad \Delta = \alpha\epsilon - \delta\gamma$$

$$\beta = \sum v^4 \qquad\qquad K_m^0 = \frac{\beta\gamma - \alpha\delta}{\Delta}$$

$$\gamma = \sum \frac{v^3}{S} \qquad\qquad V_{max}^0 = \frac{\beta\epsilon - \delta^2}{\Delta}$$

$$\delta = \sum \frac{v^4}{S}$$

$$\epsilon = \sum \frac{v^4}{S^2}$$

where K_m^0 and V_{max}^0 are the provisional estimates of K_m and V_{max}.
To obtain a fine adjustment to the parameters, the following
equations are used:

$$f = \frac{V_{max}^0 \, S_i}{S_i + K_m^0}$$

$$f' = -\frac{V_{max}^0 \, S_i}{(S_i + K_m^0)^2}$$

$$\alpha = \sum f^2$$

$$\Delta = \alpha\beta - \gamma^2$$

$$\beta = \sum f'^2$$

$$b_1 = \frac{\beta\delta - \gamma\varepsilon}{\Delta}$$

$$\gamma = \sum ff'$$

$$b_2 = \frac{\alpha\varepsilon - \gamma\delta}{\Delta}$$

$$\delta = \sum vf$$

$$\varepsilon = \sum vf'$$

$$V_{max} = b_1 V_{max}^0$$

$$K_m = K_m^0 + b_2/b_1$$

Standard error of K_m (S.E.) = $\dfrac{S}{b_1\sqrt{\alpha/\Delta}}$

where

$$S = \sqrt{(\Sigma v^2 - b_1 - b_2)/(n - 2)}$$

Standard error of V_{max} (S.E.) = $V_{max}^0 S\sqrt{\beta/\Delta}$

The equations were programmed on the HP-67 and TI-59 calculators, and the keystroke listings and user instructions are provided below.

User Instructions--RPN

Step	Instructions	Input	Keys	Output
1	Load program			
2	Initialize		f e	0.000
3	Enter v_i	v_i	ENTER↑	v_i
4	Enter S_i	S_i	A	n
5	Compute provisional estimates of K_m^0 and V_{max}^0		B	K_m^0 V_{max}^0
	For refinement of provisional estimates data must be re-entered			
6	Enter v_i	v_i	ENTER↑	v_i
7	Enter S_i	S_i	C	n
8	Compute refined K_m		D	K_m
9	Compute standard error of K_m		R/S	S.E.
10	Compute refined V_{max}		E	V_{max}
11	Compute standard error of V_{max}		R/S	S.E.
	For Michaelis-Menten plot (v vs. S):			
12	Enter S		f a	v
	For Lineweaver-Burk plot ($1/v$ vs. $1/S$):			
13	Enter $1/S$		f b	$1/v$

Program Listing--RPN

Line	Key	Comments
001	*LBLA	Data entry
002	STO1	→ last S_i
003	R↓	
004	STO0	→ last v_i
005	X²	
006	X²	
007	ST+3	→ $\sum v^4$
008	RCL1	
009	÷	
010	ST+5	→ $\sum v^4/S$
011	RCL0	
012	3	
013	Yˣ	
014	ST+2	→ $\sum v^3$
015	RCL1	
016	÷	
017	ST+4	→ $\sum v^3/S$
018	RCL0	
019	X²	
020	X²	
021	RCL1	
022	X²	
023	÷	
024	ST+6	→ $\sum v^4/S^2$
025	1	
026	ST+9	→ n
027	RCL9	
028	RTN	
029	*LBLB	
030	RCL2	Compute K_m^0 and V_{max}^0
031	RCL6	
032	x	
033	RCL4	
034	RCL5	
035	x	
036	-	
037	STOE	→ Δ
038	RCL3	
039	RCL4	
040	x	
041	RCL2	
042	RCL5	
043	x	
044	-	
045	RCLE	

Line	Key	Comments
046	÷	
047	STOA	→ K_m^0
048	RCL3	
049	RCL6	
050	x	
051	RCL5	
052	X²	
053	-	
054	RCLE	
055	÷	
056	STOB	→ V_{max}^0
057	RCLA	
058	PRTX	Display K_m^0 and V_{max}^0
059	RCLB	
060	PRTX	
061	SPC	
062	P≠S	
063	RTN	
064	*LBLC	Data reentry
065	STO1	
066	STO2	→ last S_i
067	R↓	
068	STO0	→ last v_i
069	X²	→ $\sum v^2$
070	ST+8	
071	RCLA	
072	ST+2	→ $S_i + K_m^0$
073	RCLB	
074	RCL1	
075	x	
076	RCL2	
077	÷	
078	ENT↑	
079	X²	→ $\sum f^2$
080	ST+3	
081	LSTX	
082	RCL0	
083	x	
084	ST+6	→ $\sum vf$
085	R↓	
086	R↓	
087	RCLB	
088	RCL1	
089	x	
090	CHS	

Line	Key	Comments
091	RCL2	
092	X^2	
093	÷	
094	X^2	
095	ST+4	→ $\sum f'^2$
096	LSTX	
097	X⇄Y	
098	CLX	
099	R↓	
100	ENT↑	
101	ENT↑	
102	RCL0	
103	X	
104	ST+7	→ $\sum vf'$
105	R↓	
106	X	
107	ST+5	→ $\sum ff'$
108	1	
109	ST+9	→ n
110	RCL9	
111	RTN	
112	*LBLD	
113	RCL3	Compute refined K_m
114	RCL4	and standard error
115	X	of K_m
116	RCL5	
117	X^2	
118	-	
119	STOE	→ Δ
120	RCL4	
121	RCL6	
122	X	
123	RCL5	
124	RCL7	
125	X	
126	-	
127	RCLE	
128	÷	
129	STOC	→ b_1
130	RCL3	
131	RCL7	
132	X	
133	RCL5	
134	RCL6	
135	X	
136	-	
137	RCLE	

Line	Key	Comments
138	÷	
139	STOD	→ b_2
140	RCLC	
141	RCLB	
142	X	
143	RCLB	
144	STOI	
145	R↓	
146	STOB	→ refined V_{max}
147	RCLD	
148	RCLC	
149	÷	
150	RCLA	
151	+	
152	STOA	→ refined K_m
153	RCL8	
154	RCLC	
155	RCL6	
156	X	
157	-	
158	RCLD	
159	RCL7	
160	X	
161	-	
162	RCL9	
163	2	
164	-	
165	÷	
166	$\sqrt{X}$	
167	STO2	→ S
168	RCLC	
169	÷	
170	RCL3	
171	RCLE	
172	÷	
173	$\sqrt{X}$	
174	X	
175	RCLA	
176	PRTX	
177	X⇄Y	
178	PRTX	
179	SPC	
180	RTN	
181	*LBLE	Compute and display
182	RCLI	standard error of
183	RCL2	V_{max}
184	X	

Line	Key	Comments
185	RCL4	
186	RCLE	
187	÷	
188	√X	
189	x	
190	RCLB	
191	PRTX	
192	X≠Y	
193	PRTX	
194	SPC	
195	RTN	
196	*LBLα	
197	STO0	Selection of
198	PRTX	Michaelis-Menten
199	RCLB	plot?
200	x	
201	RCL0	
202	RCLA	
203	+	
204	÷	
205	PRTX	
206	SPC	
207	RTN	
208	*LBLb	Selection of
209	PRTX	Lineweaver-Burk
210	RCLA	plot?
211	RCLB	
212	÷	
213	x	
214	RCLB	
215	1/X	
216	+	
217	PRTX	
218	SPC	
219	RTN	
220	*LBLe	Initialization
221	CLRG	
222	P≠S	
223	CLRG	
224	RTN	

Register Contents, Labels, and Data Cards--RPN

Register	Contents	Register	Contents	Labels	Contents
R_0	Last v_i	A	K_m	A	$v_i \uparrow S_i$
R_1	Last S_i	B	V_{max}	B	$\rightarrow K_m^0, V_{max}^0$
R_2	$\sum v^3$	C	b_1	C	$v_i \uparrow S_i$
R_3	$\sum v^4$	D	b_2	D	$\rightarrow K_m$; S.E.
R_4	$\sum v^3/S$	E	Δ	E	$\rightarrow V_{max}$; S.E.
R_5	$\sum v^4/S$			a	M-M plot?
R_6	$\sum v^4/S^2$			b	L-B plot?
R_9	n			c	Used
R_{s0}	Last v_i				
R_{s1}	Last S_i				
R_{s2}	$S_i^0 + K_m^0$				
R_{s3}	$\sum f^2$				
R_{s4}	$\sum f'^2$				
R_{s5}	$\sum ff'$				
R_{s6}	$\sum vf$				
R_{s7}	$\sum vf'$				
R_{s8}	$\sum v^2$				
R_{s9}	n				

User Instructions--Algebraic System--Part I

Step	Instructions	Input	Keys	Output
1	Print toggle		2nd B'	
2	Enter velocity	v	A	v
3	Enter substrate concentration and all data pairs v,S	S	R/S	S
4	Calculate K_m^0		B	K_m^0
5	Calculate v_{max}^0		R/S	v_{max}^0

*The program is divided into two parts. The first routine gives extimates of K_m^0 and v_{max}^0.

Program Listing--Algebraic System--Part I

Line	Key	Entry	Comments	Line	Key	Entry	Comments
000	76	LBL		023	85	+	
001	17	B'	Print toggle	024	43	RCL	
002	87	IFF		025	29	29	
003	08	08		026	91	R/S	
004	45	Y×		027	42	STD	Enter S
005	86	STF		028	01	01	
006	08	08		029	42	STD	
007	92	RTN		030	29	29	
008	76	LBL		031	02	2	
009	45	Y×		032	01	1	Call print
010	22	INV		033	71	SBR	subroutine
011	86	STF		034	85	+	
012	08	08		035	43	RCL	
013	92	RTN		036	00	00	
014	76	LBL	Enter v	037	33	X²	
015	11	A		038	33	X²	
016	42	STD		039	44	SUM	
017	00	00		040	03	03	$\sum v^4$
018	42	STD	Call print	041	55	÷	
019	29	29	subroutine	042	43	RCL	
020	02	2		043	01	01	
021	00	0		044	95	=	
022	71	SBR		045	44	SUM	$\sum v^4/S$

Line	Key	Entry	Comments		Line	Key	Entry	Comments
046	05	05			089	04	04	
047	43	RCL			090	65	×	
048	00	00			091	43	RCL	
049	45	Y×			092	05	05	
050	03	3			093	54	)	
051	95	=			094	95	=	
052	44	SUM			095	42	STO	
053	02	02			096	08	08	
054	55	÷	$\sum v^3$		097	53	(	
055	43	RCL			098	53	(	
056	01	01			099	43	RCL	
057	95	=			100	03	03	
058	44	SUM			101	65	×	
059	04	04			102	43	RCL	
060	43	RCL	$\sum v^3/S$		103	04	04	
061	00	00			104	54	)	
062	33	X²			105	75	-	
063	33	X²			106	53	(	
064	55	÷			107	43	RCL	
065	43	RCL			108	02	02	
066	01	01			109	65	×	
067	33	X²			110	43	RCL	
068	95	=			111	05	05	
069	44	SUM			112	54	)	
070	06	06			113	54	)	
071	01	1	$\sum v^4/S^2$		114	55	÷	
072	44	SUM			115	43	RCL	
073	07	07			116	08	08	
074	43	RCL			117	95	=	
075	07	07			118	42	STO	
076	91	R/S			119	09	09	
077	76	LBL			120	42	STO	
078	12	B			121	29	29	
079	53	(	K_m^0		122	02	2	
080	43	RCL			123	02	2	
081	02	02			124	71	SBR	
082	65	×			125	85	+	Call print
083	43	RCL			126	91	R/S	subroutine
084	06	06			127	53	(	
085	54	)			128	53	(	v_{max}^0
086	75	-			129	43	RCL	
087	53	(			130	03	03	
088	43	RCL			131	65	×	

Line	Key	Entry	Comments	Line	Key	Entry	Comments
132	43	RCL		173	69	OP	
133	06	06		174	00	00	
134	54	)		175	98	ADV	
135	75	-		176	92	RTN	
136	43	RCL					
137	05	05					
138	33	X²					
139	54	)					
140	55	÷					
141	43	RCL					
142	08	08					
143	95	=					
144	42	STO					
145	10	10					
146	42	STO					
147	29	29					
148	02	2					
149	03	3					
150	71	SBR					
151	85	+	Call print				
152	91	R/S	subroutine				
153	76	LBL					
154	85	+					
155	87	IFF					
156	08	08					
157	95	=					
158	43	RCL					
159	29	29					
160	92	RTN					
161	76	LBL					
162	95	=					
163	42	STO					
164	25	25					
165	73	RC*					
166	25	25	Print				
167	69	OP	subroutine				
168	04	04					
169	43	RCL					
170	29	29					
171	69	OP					
172	06	06					

Register Contents, Labels, and Data Cards--Algebraic System, Part I

Register	Contents	Labels	Contents
R0	v	Label B'	Print toggle
R1	s	Label A	Enter v and S
R2	$\sum v^3$	Label B	Calculate K_m^0
R3	$\sum v^4$		
R4	$\sum v^3/S$		
R5	$\sum v^4/S$		
R6	$\sum v^4/S^2$		
$7	n		
R8	Δ		
R9	K_m^0		
R10	v_{max}^0		

Data Card (4 2nd Write)

Alpha Print Code	Register	Contents
42000000.	20	V
36000000.	21	S
26306500.	22	KM =
42301344.	23	VMAX

User Instructions--Algebraic System, Part II*

Step	Instructions	Input	Keys	Output
1	Enter substrate concentrations	S	C	S
2	Enter velocity and all data pairs S, v	v	R/S	n
3	Calculate V_{max}		D	V_{max}
4	Calculate K_m		R/S	K_m
5	Calculate standard error for K_m		R/S	S.E. of K_m
6	Calculate standard error for V_{max}		E	S.E. of V_{max}
7	Enter S for projection of Lineweaver-Burk	S	2nd E'	$1/v$

*The second routine requires TI-58 users to repartition the cal-
culator into 319 programming steps and 19 data registers (i.e.,
Key 2 2nd Op. 17). Note that plot projection is for TI-59 only.

Program Listing--Algebraic System, Part II

Line	Key	Entry	Comments	Line	Key	Entry	Comments
000	76	LBL		041	54	)	
001	16	A'	v_{max}^0	042	55	÷	
002	42	STD		043	53	(	
003	14	14		044	53	(	
004	91	R/S		045	43	RCL	
005	76	LBL		046	00	00	
006	17	B'	K_m^0	047	85	+	
007	42	STD		048	43	RCL	
008	15	15		049	15	15	
009	91	R/S		050	54	)	
010	76	LBL		051	33	X²	
011	13	C		052	54	)	
012	42	STD		053	95	=	f'
013	00	00	Enter S	054	94	+/-	
014	91	R/S		055	42	STD	
015	42	STD	Enter v	056	03	03	
016	01	01		057	43	RCL	
017	53	(		058	02	02	
018	43	RCL		059	33	X²	
019	14	14		060	95	=	$\alpha = \sum f^2$
020	65	×		061	44	SUM	
021	43	RCL		062	04	04	
022	00	00		063	43	RCL	
023	54	)		064	03	03	
024	55	÷		065	33	X²	$\beta = \sum f'^2$
025	53	(		066	95	=	
026	43	RCL		067	44	SUM	
027	00	00		068	05	05	
028	85	+		069	43	RCL	
029	43	RCL		070	02	02	
030	15	15		071	65	×	
031	54	)		072	43	RCL	
032	95	=		073	03	03	$\delta = \sum ff'$
033	42	STD	f	074	95	=	
034	02	02		075	44	SUM	
035	53	(		076	06	06	
036	43	RCL		077	43	RCL	
037	14	14		078	01	01	
038	65	×		079	65	×	
039	43	RCL		080	43	RCL	
040	00	00		081	02	02	

Line	Key	Entry	Comments	Line	Key	Entry	Comments
082	95	=		123	75	-	
083	44	SUM		124	53	(	
084	07	07		125	43	RCL	
085	43	RCL		126	06	06	
086	03	03		127	65	×	
087	65	×		128	43	RCL	
088	43	RCL		129	08	08	
089	01	01		130	54	)	
090	95	=	E	131	54	)	
091	44	SUM		132	55	÷	
092	08	08		133	43	RCL	
093	43	RCL		134	09	09	
094	01	01		135	95	=	
095	33	X²	$\sum v^2$	136	42	STO	b_1
096	95	=		137	10	10	
097	44	SUM		138	53	(	
098	18	18		139	53	(	
099	53	(		140	43	RCL	
100	43	RCL		141	04	04	
101	04	04		142	65	×	
102	65	×		143	43	RCL	
103	43	RCL		144	08	08	
104	05	05		145	54	)	
105	54	)		146	75	-	
106	75	-		147	53	(	
107	53	(		148	43	RCL	
108	43	RCL		149	06	06	
109	06	06		150	65	×	
110	33	X²		151	43	RCL	
111	54	)	Δ	152	07	07	
112	95	=		153	54	)	
113	42	STO		154	54	)	
114	09	09		155	55	÷	
115	53	(		156	43	RCL	
116	53	(		157	09	09	b_2
117	43	RCL		158	95	=	
118	05	05		159	42	STO	
119	65	×		160	11	11	
120	43	RCL		161	01	1	
121	07	07		162	44	SUM	
122	54	)		163	16	16	

Line	Key	Entry	Comments	Line	Key	Entry	Comments
164	43	RCL		205	65	×	
165	16	16	n	206	43	RCL	
166	91	R/S		207	07	07	
167	76	LBL		208	95	=	
168	14	D		209	42	STO	
169	43	RCL		210	00	00	
170	10	10		211	43	RCL	
171	65	×		212	11	11	
172	43	RCL		213	65	×	
173	14	14		214	43	RCL	
174	95	=	V_{max}	215	08	08	
175	42	STO		216	95	=	
176	12	12		217	42	STO	
177	91	R/S		218	01	01	
178	43	RCL		219	53	(	
179	15	15		220	43	RCL	
180	85	+		221	18	18	
181	53	(		222	75	-	
182	43	RCL		223	43	RCL	
183	11	11		224	00	00	
184	55	÷		225	75	-	
185	43	RCL		226	43	RCL	
186	10	10		227	01	01	
187	54	)	K_m	228	54	)	
188	95	=		229	55	÷	
189	42	STO		230	43	RCL	
190	13	13		231	20	20	
191	91	R/S		232	95	=	
192	02	2	$(n - 2)$	233	34	√X	S
193	42	STO		234	42	STO	
194	02	02		235	03	03	
195	43	RCL		236	53	(	
196	16	16		237	43	RCL	
197	75	-		238	03	03	
198	43	RCL		239	55	÷	
199	02	02		240	43	RCL	
200	95	=		241	10	10	
201	42	STO		242	54	)	
202	20	20		243	65	×	
203	43	RCL		244	53	(	
204	10	10					

Line	Key	Entry	Comments	Line	Key	Entry	Comments
245	53	(		287	65	×	
246	43	RCL		288	53	(	
247	04	04		289	43	RCL	
248	55	÷		290	13	13	
249	43	RCL		291	55	÷	
250	09	09		292	43	RCL	
251	54	)		293	12	12	
252	34	ΓX		294	54	)	
253	54	)	S.E. of K_m	295	85	+	
254	95	=		296	53	(	
255	42	STD		297	43	RCL	
256	19	19		298	12	12	
257	91	R/S		299	35	1/X	
258	76	LBL	S.E. of V_{max}	300	54	)	1/v
259	15	E		301	95	=	
260	53	(		302	91	R/S	
261	43	RCL					
262	14	14					
263	65	×					
264	43	RCL					
265	03	03					
266	54	)					
267	65	×					
268	53	(					
269	53	(					
270	43	RCL					
271	05	05					
272	55	÷					
273	43	RCL					
274	09	09					
275	54	)					
276	34	ΓX					
277	54	)					
278	95	=	Lineweaver-				
279	42	STD	Burk projec-				
280	24	24	tion: Enter				
281	91	R/S	S				
282	76	LBL					
283	10	E'					
284	35	1/X					
285	42	STD					
286	25	25					

Register Contents, Labels, and Data Cards--Algebraic System, Part II

Register	Contents	Labels	Contents
R0	S	Label A'	V_{max}^0
R1	v	Label C	S
R2	f	Label D	V_{max}
R3	f'	Label E	S.E. of V_{max}
R4	α		
R5	β		
R6	δ		
R7	g		
R8	E		
R9	Δ		
R10	b_1		
R11	b_2		
R12	V_{max}		
R13	K_m		
R14	V_{max}^0		
R15	K_m^0		
R16	n		
R17	S		
R18	$\sum v^2$		
R19	S.E. of K_m		

Note that several registers are written over (used twice) in this program.

Example

Below are the initial velocity data for the enzyme nicotinamide mononucleotide adenylyltransferase from the paper by
Wilkinson (2).   These data are plotted in Figure 4D1 as velocity
versus substrate concentration.   The double reciprocal Lineweaver-
Burk plot is shown in Figure 4D2.

Calculate initial and refined estimates of K_m and V_{max} and
the standard errors for each.

S^*	$v^\dagger$
0.138	0.148
0.220	0.171
0.291	0.234
0.560	0.324
0.766	0.390
1.460	0.493

Figure 4D1.   Weighted least squares--plot of velocity versus
substrate concentration.

*Concentration of nicotinamide mononucleotide, (mM).
$\dagger$Micromoles of nicotinamide adenine dinucleotide formed per mil-
ligram of enzyme protein in 3 min.

Solution:

$$K_m^0 \quad = 0.571$$
$$V_{max}^0 = 0.680$$
$$K_m \quad = 0.595; \text{ S.E. } = 0.064$$
$$V_{max} = 0.690; \text{ S.E. } 0.035$$

Lineweaver-Burk plot of fitted line

$1/S$	$1/v$
0.5	1.881
1.0	2.312
1.5	2.743
2.0	3.174
3.0	4.037
5.0	5.761

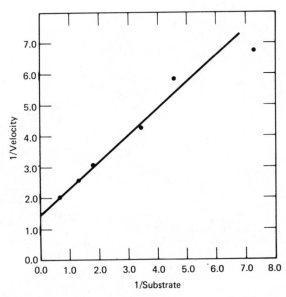

Figure 4D2. Weighted least squares--plot of $1/v$ versus $1/S$.

References

1. W.W. Cleland (1967), "The Statistical Analysis of Enzyme Kinetic Data," *Advances in Enzymology 29*, 1-32.

2. G.N. Wilkinson (1961), "Statistical Estimations in Enzyme Kinetics," *Biochemical Journal 80*, 324-332.

3. M.E. Magar (1972), *Data Analysis in Biochemistry and Biophysics*, Academic Press, New York, pp. 429-432.

4. M.R. Atkinson, J.F. Jackson, and R.K. Morton (1961), *Biochemical Journal 80*, 318-323.

4E. DETERMINATION OF K_m AND V_{max} FROM A SINGLE PROGRESS CURVE

In this section we present an algorithm and a program for determining the Michaelis kinetic parameters, K_m and V_{max}, for an enzyme catalyzed reaction by analysis of a total reaction progress curve. In principle, one estimates the velocities at several substrate concentrations throughout the course of a complete reaction. These velocities can then be treated by the weighted linear regression method described in the preceding section.

For the integrated Michaelis-Menten equation:

$$V_{max}t = (S_0 - S_t) + K_m \ln \frac{S_0}{S_t} \tag{1}$$

there exist pairs of data (t_i, S_i) such that

$$Vt_i = S_0 - S_i + K_m \ln \frac{S_0}{S_i} \qquad\qquad Vt_j = S_0 - S_j + K_m \ln \frac{S_0}{S_j}$$

from which one obtains by subtraction

$$- V_{max}(t_i - t_j) = S_i - S_j + K_m \ln \frac{S_i}{S_j} \tag{2}$$

By inverting and rearranging

$$\frac{1}{v_i} = \frac{1}{V_{max}} + \frac{K_m}{V_{max}} \frac{\ln (S_i/S_j)}{S_i - S_j} \tag{3}$$

If we let $S = (S_i - S_j)/\ln (S_i/S_j)$

$$\frac{1}{v_i} = \frac{1}{V_{max}} + \frac{K_m}{V_{max}S} \tag{4}$$

In this form, then, the data can be entered into the weighted linear regression routine used in the preceding section. The terms S_i and S_j represent substrate concentrations at two points along a reaction progress curve separated by the time interval Δt. It is not necessary to use fixed intervals of ΔS and Δt.

The recommended procedure for obtaining data pairs S_i, t_i and S_j, t_j is outlined in Figure 4E.

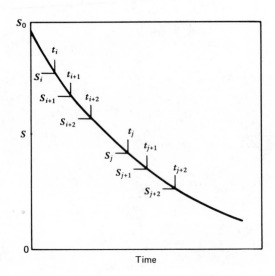

Figure 4E. Data selection for single progress curve

User Instructions--RPN

Step	Instructions	Input	Keys	Output
1	Load program and initialize		f e	0.00
2	If reaction is measured by following absorbance of product, enter extinction coefficient of product	$\varepsilon(P)$	f a	$\varepsilon(P)$
3	If reaction is measured by following absorbance of substrate, enter extinction coefficient of substrate	$\varepsilon(S)$	f b	$\varepsilon(S)$
4	If step 2 is followed, you <u>must</u> enter initial substrate concentration, S_0	S_0	f c	S_0
5	Enter absorbance, A_i, at time t_i	A_i	ENTER↑	
6	Enter time t_i	t_i	A	A_i
7	Enter absorbance, A_j, at time t_j	A_j	ENTER↑	
8	Enter time t_j	t_j	B	$\overline{v}$
9	If substrate concentration is known at each time point, simply enter S_i, t_i and S_j, t_j at steps 5-8			
10	To compute and display $\overline{S}$		R/S	$\overline{S}$
11	Accumulate sums		R/S	n
12	For automatic output of $\overline{v}$ and $\overline{S}$ on printer, that is, steps 10 and 11 are automatic		f d	
13	Calculate K_m		C	K_m
14	Calculate V_{max}		D	V_{max}

Program Listing--RPN

Line	Key	Comments	Line	Key	Comments
001	*LBLa		046	X^2	
002	STO6	Enter extinction	047	ENT↑	
003	SF0	coefficient for	048	ENT↑	
004	RTN	product	049	ST+1	
005	*LBLb		050	RCLI	
006	STO6	Enter extinction	051	÷	
007	SF1	coefficient for	052	ST+3	Accumulate sums for
008	RTN	substrate	053	X⇄Y	weighted regression
009	*LBLA		054	RCLI	
010	STOB	→ t_i	055	X^2	
011	R↓		056	÷	
012	STOA	→ A_i	057	ST+4	
013	RTN		058	RCLE	
014	*LBLB		059	3	
015	STOD	→ t_j	060	Y^x	
016	R↓		061	ST+0	
017	STOC	→ A_j	062	RCLI	
018	F0?		063	÷	
019	GSB0	Compute substrate	064	ST+2	
020	F1?	concentration from	065	1	
021	GSB1	absorbance of pro-	066	ST+5	
022	RCLA	duct or substrate	067	RCL5	
023	RCLC		068	RTN	
024	-		069	*LBLC	
025	ENT↑		070	RCL0	
026	ENT↑		071	RCL4	
027	RCLD		072	x	
028	RCLB		073	RCL2	
029	-		074	RCL3	
030	ENT↑		075	x	Compute K_m and store
031	R↓	Compute $\bar{v}$	076	-	in R_A
032	÷		077	STOE	
033	STOE	$\bar{v} = \Delta A / \Delta t$	078	RCL1	
034	GSB9		079	RCL2	
035	X⇄Y	Output $\bar{v}$	080	x	
036	RCLA		081	RCL0	
037	RCLC	Compute $\bar{S}$	082	RCL3	
038	÷		083	x	
039	LN	$\bar{S} = \dfrac{(S_i - S_j)}{\ln(S_i/S_j)}$	084	-	
040	÷		085	RCLE	
041	STOI		086	÷	
042	GSB9	Output $\bar{S}$	087	STOA	
043	SPC		088	PRTX	
044	RCLE		089	RTN	
045	X^2		090	*LBLD	

Line	Key	Comments	Line	Key	Comments
091	RCL1		138	RTN	
092	RCL4	Compute V_{max} and	139	*LBLd	Print flag: Set and
093	x	store in R_B	140	0	Reset
094	RCL3		141	F2?	
095	X²		142	RTN	
096	-		143	1	
097	RCLE		144	SF2	
098	÷		145	RTN	
099	STOB		146	*LBL9	Print subroutine
100	PRTX		147	F2?	
101	SPC		148	GTO8	
102	RTN		149	R/S	
103	*LBL0	Compute S_i and S_j	150	RTN	
104	RCLA	from absorbance of	151	*LBL8	
105	RCL6	product	152	PRTX	
106	÷		153	SF2	
107	RCL7		154	RTN	
108	X⇄Y				
109	-				
110	STOA				
111	RCLC				
112	RCL6				
113	÷				
114	RCL7				
115	X⇄Y				
116	-				
117	STOC				
118	RTN				
119	*LBL1	Compute S_i and S_j			
120	RCLA	from absorbance of			
121	RCL6	substrate			
122	÷				
123	STOA				
124	RCLC				
125	RCL6				
126	÷				
127	STOC				
128	RTN				
129	*LBLc	Store initial sub-			
130	STO7	strate concentration			
131	RTN	S_0			
132	*LBLe	Initialization			
133	CLRG				
134	P⇄S				
135	CLRG				
136	CF0				
137	CF1				

Register Contents, Labels, and Data Cards--RPN

Register	Contents	Labels	Contents
R_0	$\sum v^3$	A	$A_i \uparrow t_i$
R_1	$\sum v^4$	B	$A_j \uparrow t_j$
R_2	$\sum v^3/S$	C	$\rightarrow K_m$
R_3	$\sum v^4/S$	D	$\rightarrow V_{max}$
R_4	$\sum v^4/S^2$	a	$\varepsilon(P)$
R_5	n	b	$\varepsilon(S)$
R_6	ε	c	S_0
R_7	S_0	d	Print?
R_A	Last A_i (K_m)	e	Initialize
R_B	Last t_i (V_{max})		
R_C	Last A_j		
R_D	Last t_j		
R_E	Last $\bar{v}(\Delta)$		
R_I	Last $\bar{S}$		

User Instructions--Algebraic System

Step	Instructions	Input	Keys	Output
1	If reaction is measured by following change in optical density of product, enter extinction coefficient	ε_p	2nd A'	
2	If reacion is measured by following change in optical density of substrate, enter extinction coefficient	ε_s	2nd B'	
3	If step 1 is followed, enter initial substrate concentration	S_0	2nd C'	
4	Enter absorbance i (or S_i^* if known) and time i	A_i t_i	A R/S	
5	Enter absorbance j (or S_j^* if known) and time	A_j t_j	B R/S	v_i
6	Output v_i		R/S	S_i
7	Output S_i		R/S	n
	Repeat steps 4-7 for all data			
8	Calculate K_m		C	K_m
9	Calculate V_{max}		R/S	V_{max}

*If the substrate concentration is known at each point in time, enter S_i, t_i at step 4 and S_j, t_j at step 5.

Program Listing--Algebraic System

Line	Key	Entry	Comments	Line	Key	Entry	Comments
000	76	LBL		041	43	RCL	
001	16	A'	Enter ϵ_p	042	12	12	
002	86	STF		043	75	-	
003	00	00		044	43	RCL	
004	42	STD		045	14	14	
005	18	18		046	54	)	
006	92	RTN		047	42	STD	
007	76	LBL	Enter ϵ_s	048	23	23	
008	17	B'		049	55	÷	
009	86	STF		050	53	(	
010	01	01		051	43	RCL	
011	42	STD		052	15	15	
012	18	18		053	75	-	
013	76	LBL		054	43	RCL	
014	18	C'	Enter initial	055	13	13	
015	42	STD	substrate	056	95	=	
016	20	20		057	42	STD	
017	92	RTN		058	00	00	$\bar{v}$
018	76	LBL	Enter S_i	059	91	R/S	
019	11	A		060	43	RCL	
020	42	STD		061	23	23	
021	12	12		062	55	÷	
022	91	R/S	Enter t_i	063	53	(	
023	42	STD		064	43	RCL	
024	13	13		065	12	12	
025	92	RTN		066	55	÷	
026	76	LBL	Enter S_j	067	43	RCL	
027	12	B		068	14	14	
028	42	STD		069	54	)	
029	14	14		070	23	LNX	
030	91	R/S	Enter t_j	071	95	=	
031	42	STD		072	42	STD	$\bar{s}$
032	15	15		073	01	01	
033	87	IFF		074	92	RTN	
034	89	89		075	43	RCL	
035	87	IFF		076	00	00	
036	01	1		077	33	X²	
037	38	SIN		078	33	X²	
038	76	LBL		079	44	SUM	
039	30	TAN		080	03	03	$\sum v^4$
040	53	(		081	55	÷	

Line	Key	Entry	Comments	Line	Key	Entry	Comments
082	43	RCL		123	43	RCL	
083	01	01		124	06	06	
084	95	=		125	54	)	
085	44	SUM	$\sum v^4/s$	126	75	-	
086	05	05		127	53	(	
087	43	RCL		128	43	RCL	
088	00	00		129	04	04	
089	45	Y^X		130	65	×	
090	03	3		131	43	RCL	
091	95	=		132	05	05	
092	44	SUM	$\sum v^3$	133	54	)	
093	02	02		134	95	=	
094	55	÷		135	42	STO	
095	43	RCL		136	08	08	
096	01	01		137	53	(	
097	95	=		138	53	(	
098	44	SUM	$\sum v^3/s$	139	43	RCL	
099	04	04		140	03	03	
100	43	RCL		141	65	×	
101	00	00		142	43	RCL	
102	33	X^2		143	04	04	
103	33	X^2		144	54	)	
104	55	÷		145	75	-	
105	43	RCL		146	53	(	
106	01	01		147	43	RCL	
107	33	X^2		148	02	02	
108	95	=		149	65	×	
109	44	SUM	$\sum v^4/s^2$	150	43	RCL	
110	06	06		151	05	05	
111	01	1		152	54	)	
112	44	SUM		153	54	)	
113	07	07	n	154	55	÷	
114	43	RCL		155	43	RCL	
115	07	07		156	08	08	
116	91	R/S		157	95	=	
117	76	LBL	K_m	158	42	STO	
118	13	C		159	09	09	
119	53	(		160	91	R/S	V_{max}
120	43	RCL		161	53	(	
121	02	02		162	53	(	
122	65	×		163	43	RCL	

Line	Key	Entry	Comments	Line	Key	Entry	Comments
164	03	03		205	20	20	
165	65	×		206	75	-	
166	43	RCL		207	53	(	
167	06	06		208	43	RCL	
168	54	)		209	12	12	
169	75	-		210	55	÷	
170	43	RCL		211	43	RCL	
171	05	05		212	18	18	
172	33	X²		213	54	)	
173	54	)		214	95	=	
174	55	÷		215	42	STD	
175	43	RCL		216	12	12	
176	08	08		217	43	RCL	
177	95	=		218	20	20	
178	42	STD		219	75	-	
179	10	10		220	53	(	
180	91	R/S		221	43	RCL	
181	76	LBL	Subroutine	222	14	14	
182	38	SIN	for calcula-	223	55	÷	
183	43	RCL	ting sub-	224	43	RCL	
184	12	12	strate con-	225	18	18	
185	55	÷	centration	226	54	)	
186	43	RCL	from input of	227	95	=	
187	18	18	product opti-	228	42	STD	
188	95	=	cal density	229	14	14	
189	42	STD		230	71	SBR	
190	12	12		231	30	TAN	
191	43	RCL		232	92	RTN	
192	14	14					
193	55	÷					
194	43	RCL					
195	18	18					
196	95	=					
197	42	STD					
198	14	14					
199	71	SBR					
200	30	TAN					
201	92	RTN					
202	76	LBL	Subroutine				
203	89	π	for calcula-				
204	43	RCL	ting sub-				
			strate con-				
			centration				
			from O.D.				

Register Contents, Labels, and Data Cards--Algebraic System

Register	Contents	Labels	Contents
R0	$\bar{v}$	Label A'	ε_p
R1	$\bar{s}$	Label B'	ε_s
R2	$\sum v^3$	Label C'	s_0
R3	$\sum v^4$	Label A	A_i
R4	$\sum v^3/s$	Label B	A_j
R5	$\sum v^4/s$	Label C	K_m
R6	$\sum v^4/s^2$		
R7	n		
R8	K_m		
R9	V_{max}		
R12	A_i		
R13	t_i		
R14	A_j		
R15	t_j		
R18	ε_p		
R20	s_0		
R23	$(s_i - s_j)$		

Example

Below are data from a complete reaction progress curve for the hydrolysis of phenyl phosphate catalyzed by prostate acid phosphatase, previously presented by Schonheyder (1) and used by Yun and Suelter (2) to demonstrate the single progress curve method. Only odd-numbered sets of data were used in this example. Initial substrate concentration (S_0) is 0.9838 mM.

	P (mM)	S_i (mM)	t_i (min)	S_j (mM)	t_j (min)	$\bar{V}$	$\bar{S}$
1	0.0252	0.9586	0.47	0.7694	4.57	0.0461	0.8605
2	0.1135	0.8703	2.35	0.6685	6.9	0.0444	0.7650
3	0.2144	0.7694	4.57	0.5677	9.7	0.0393	0.6634
4	0.3153	0.6685	6.9	0.4920	12.03	0.0344	0.5757
5	0.4161	0.5677	9.7	0.4163	14.88	0.0292	0.4881
6	0.4918	0.4920	12.03	0.3407	17.85	0.0260	0.4117
7	0.5675	0.4163	14.88	0.2650	21.51	0.0228	0.3350
8	0.6431	0.3407	17.85	0.1894	26.78	0.0169	0.2577
9	0.7188	0.2650	21.51	0.1263	32.33	0.0128	0.1872
10	0.7944	0.1894	26.78	0.0759	39.4	0.0090	0.1241
11	0.8575	0.1263	32.33	0.0254	54.9	0.0045	0.0629
12	0.9079	0.0759	39.4				
13	0.9584	0.0254	54.9				

Solution

K_m = 2.3634 mM
V_{max} = 0.1768 mole/min·mg protein

References

1. F. Schonheyder (1952), *Biochemical Journal 50*, 378-384.

2. S.-L. Yun and C.H. Suelter (1977), *Biochemica et Biophysica Acta 480*, 1-13.

4F. ACTIVATION ENERGY--ARRHENIUS EQUATION

Estimation of the Activation Energy of a Biochemical Reaction

The Arrhenius equation is an empirical equation which relates the rate constant of a reaction to the absolute temperature:

$$k = Ae^{-E_a/RT} \tag{1}$$

Most chemical and biochemical reactions require that a certain
amount of energy be expended before initiation (i.e., there is an
energy barrier, and the reaction cannot proceed spontaneously
(Figure 4F1). The Arrhenius constant or collision frequency A
also contains a steric factor since collision of molecules alone
will not produce a reaction without proper orientation (these
terms are not strictly applicable to molecules in solution. The
$e^{-E_a/RT}$ in equation 1 represents the fraction of molecules
involved in collisions with other molecules which have energy
sufficient for the reaction to proceed.

Since the magnitude of the activation energy (E_a) is largely
responsible for the reaction rate, it is desirable to estimate
this term accurately. The problem can be approached experimentally by measuring the velocity of the enzymatic reaction (velocity is taken as an index of the rate constant k) as a function of
temperature. The Arrhenius equation is linearized (equation 2),
converted to the natural ln format and solved graphically by plotting ln k versus $1/T$ (Figure 4F2). Then E_a can be derived from
the slope of the resulting straight line.

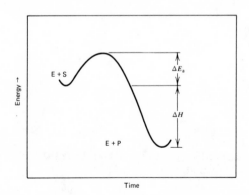

Figure 4F1. Energy profile for the following single step
            reaction: Enzyme + Substrate → Enzyme + Product

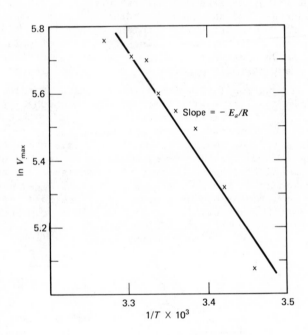

Figure 4F2.   Arrhenius plot of mitochondrial succinate-cytochrome
            c reductase.

The calculator program allows the observed values to be plotted and compared with the derived values of the fitted line. Graphical inspection is required since the behavior of enzymes as a function of temperature is not always linear when plotted against the log of the reaction rate.

$$\ln k = \ln A - \frac{E_a}{R} \frac{1}{T}$$ (2)

In some cases Arrhenius plots may have deviations from linearity, gaps, or changes in slope which are important indications of changes in enzymatic mechanism--high or low temperature denaturation of enzymatic proteins, alteration of the enzyme's environment, which in turn affects the catalytic function (1), or changes in the affinity of the enzyme for the substrate (2). If such deviations occur, groups of points which appear to fall on a straight line can be determined by inspection. A final assignment of points may then be based on the best least squares correlation coefficient for a given set of points.

Another useful description of the behavior of enzymatic systems as a function of temperature is the so-called Q_{10} value. This factor, which is usually double for every 10°C for strictly chemical reactions, serves as a good comparative index of similar enzymes from differing systems. The temperature coefficient Q_{10} is defined as the factor by which the reaction rate changes because of a 10°C alteration in temperature:

$$Q_{10} = \frac{k_2}{k_1} \frac{[10]}{[T_2 - T_1]}$$ (3)

After estimating the activation energy for a reaction, the calculator program allows Q_{10} to be determined by utilizing the relationship between the activation energy and the temperature coefficient:

$$E_a = \frac{RT_1 T_2}{T_2 - T_1} \ln \frac{k_2}{k_1}$$ (4)

$$E_a = \frac{RT_2 T_1 \ln Q_{10}}{10}$$ (5)

User Instructions--RPN

Step	Instructions	Input	Keys	Output
1	Load program and initialize		f a	0.00
2	To select print mode		f e	1.00
3	To turn off print mode		f e	0.00
4	Enter reaction rate k and temperature T (°C)	k T	ENTER↑ A	n
	If print flag is on, ln k and $1/T$)°K) × 10^3 are paused (HP-67) or printed (HP-97)			
	If print flag is off, ln k is displayed. To display $1/T$ (°K) × 10^3 press R/S To accumulate sums, press R/S again			
5	To delete incorrect data pair, enter k and T (°C)	k T	ENTER↑ B	$n - 1$
6	Repeat step 4 for all data			
7	Compute coefficient of determination and activation energy, E_a (kcal/mole)			r
8	To display pre-exponential factor	A	f c	A
9	To plot best fit straight line: enter T (°C); $1/T$ (°K) × 10^3 is displayed To display computed ln k	T (°C)	D R/S	$1/T$ (°K) ln k
10	To compute temperature coefficient, enter T_1(°C) and T_2(°C)	T_1 T_2	ENTER↑ E	Q_{10}

Program Listing--RPN

Line	Key	Comments	Line	Key	Comments
001	*LBLA		046	RCL6	
002	STO0	→ T (°C)	047	x	
003	R↓		048	RCL9	
004	STO1	→ k → ln k	049	÷	
005	LN		050	−	
006	RCL0		051	ENT↑	
007	2		052	ENT↑	
008	7		053	RCL4	
009	3		054	X²	
010	STOE		055	RCL9	
011	+		056	÷	
012	1/X		057	RCL5	
013	EEX	→ $1/T$ (°K) x 10^3	058	X≷Y	
014	3		059	−	
015	x		060	÷	
016	CF3	Clear deletion flag	061	STOB	
017	F2?	Print?	062	x	
018	GSB9		063	RCL6	
019	Σ+	Accumulate sums	064	X²	
020	RTN		065	RCL9	
021	*LBLB		066	÷	
022	STO0	→ T	067	CHS	
023	R↓		068	RCL7	
024	STO1	→ k → ln k	069	+	
025	LN		070	÷	
026	RCL0		071	STOC	
027	RCLE		072	PRTX	→ r^2
028	+		073	RCL6	Coefficient of
029	1/X		074	RCL4	determination
030	EEX		075	RCLB	
031	3		076	x	
032	x		077	−	
033	SF3	Set deletion flag	078	RCL9	
034	F2?	Print?	079	÷	
035	GSB3	Print deletion	080	STOA	
036	F2?	indicator	081	RCLB	
037	GSB9		082	CHS	
038	Σ−	Deletion	083	1	
039	RTN		084	.	
040	*LBLC	Compute regression	085	9	
041	SPC	coefficients	086	8	
042	FIX		087	7	
043	P≷S		088	x	
044	RCL8		089	DSP3	
045	RCL4				

Line	Key	Comments	Line	Key	Comments
090	P⇄S	Output	136	RCL4	
091	PRTX	E_a (kcal/mole)	137	RCL5	
092	RTN		138	–	
093	*LBLc		139	RCL4	
094	DSP3		140	RCL5	
095	RCLA	Output	141	x	
096	e^x	preexponential	142	÷	
097	SCI	factor	143	RCLB	
098	PRTX		144	CHS	
099	RTN		145	EEX	
100	*LBLD		146	3	
101	SPC	Projection routine	147	x	
102	FIX	for plotting	148	x	
103	RCLE		149	e^x	
104	+		150	F2?	
105	1/X		151	GSB4	
106	EEX		152	SPC	
107	3	$1/T$ (°K) x 10^3	153	RTN	
108	x		154	e^x	
109	F2?		155	F2?	
110	GSB4		156	GTO4	
111	R/S		157	RTN	
112	RCLB		158	*LBLe	
113	x	Compute ln k	159	0	Print flag: Set and
114	RCLA		160	F2?	Reset
115	+		161	RTN	
116	F2?		162	1	
117	GSB4	and print if print	163	SF2	
118	SPC	flag is on	164	RTN	
119	RTN		165	*LBL9	
120	*LBLa		166	SPC	Print data subrou-
121	FIX		167	X⇄Y	tine
122	DSP3		168	PRTX	
123	CLRG		169	X⇄Y	
124	P⇄S		170	PRTX	
125	CLRG		171	SF2	
126	CLX		172	RTN	
127	RTN		173	*LBL3	
128	*LBLE		174	SPC	Print deletion
129	FIX		175	1	indicator
130	STO4	Compute Q_{10} routine	176	CHS	
131	R↓		177	PRTX	
132	STO5		178	SF2	
133	RCLE		179	R↓	
134	ST+4		180	RTN	
135	ST+5				

Line	Key	Comments
181	*LBL4	Short print routine
182	PRTX	
183	SF2	
184	RTN	

Register Contents, Labels, and Data Cards--RPN

Register	Contents	Labels	Contents
R_0	Last T (°C)	A	$k \uparrow T(+)$
R_1	Last k	B	$k \uparrow T(-)$
R_{s4}	$\sum 1/T$ (°K) $\times 10^3$	C	$\rightarrow r^2, E_a$
R_{s5}	$\sum [1/T$ (°K) $\times 10^3]^2$	D	$T \rightarrow 1/T$; ln k
R_{s6}	$\sum \ln k$	E	$T_1 \uparrow T_2 \rightarrow Q_{10}$
R_{s7}	$\sum (\ln k)^2$	a	Initialization
R_{s8}	$\sum \ln k[1/T$ (°K) $\times 10^3]$	c	$\rightarrow A$
R_{s9}	n	e	Print?
R_A	ln A		
R_B	$- E_a/R$		
R_C	r^2		
R_E	273		

User Instructions--Algebraic System

Step	Instructions	Input	Keys	Output
1	Pring toggle		2nd B'	
2	Enter T (°C)	T	A	$1/T$
3	Record $1/T \times 10^3$ for plot			
4	Enter $1/T$ into least squares		R/S	
5	Enter reaction rate	k	B	$\ln k$
6	Record $\ln k$ for plot			
7	Enter all data pairs T, k before E_a calculation		R/S	n
8	Calculate E_a (kcal/mole)		C	E_a
9	Determine correlation coefficient		D	r
	To derive fitted line:			
10	Enter $1/T'$	$1/T'$	E	$\ln k'$
11	Enter T_1 (°C) for Q_{10}	T_1	E'	
12	Enter T_2 (°C) for calculation of Q_{10}	T_2	R/S	Q_{10}

Program Listing--Algebraic System

Line	Key	Entry	Comments	Line	Key	Entry	Comments
000	76	LBL		041	57	ENG	
001	17	B'	Print	042	65	×	
002	87	IFF	toggle	043	43	RCL	
003	08	08		044	14	14	
004	45	YX		045	95	=	Call print
005	86	STF		046	42	STO	subroutine
006	08	08		047	29	29	
007	92	RTN		048	02	2	
008	76	LBL		049	01	1	
009	45	YX		050	71	SBR	
010	22	INV		051	85	+	
011	86	STF		052	91	R/S	
012	08	08		053	43	RCL	Linear
013	92	RTN		054	29	29	regression
014	76	LBL	Enter T (°C)	055	36	PGM	
015	11	A		056	01	01	
016	42	STO		057	32	X:T	
017	15	15	Call Print	058	92	RTN	Enter k
018	42	STO	subroutine	059	76	LBL	
019	29	29		060	12	B	
020	02	2		061	42	STO	
021	00	0		062	13	13	
022	71	SBR		063	42	STO	
023	85	+	Convert	064	29	29	Call print
024	02	2	T (°C) to	065	02	2	subroutine
025	07	7	$1/T$ (°K)	066	02	2	
026	03	3		067	71	SBR	
027	42	STO		068	85	+	
028	16	16		069	43	RCL	
029	01	1		070	29	29	
030	52	EE		071	23	LNX	
031	03	3		072	42	STO	Convert to ln
032	42	STO		073	29	29	
033	14	14		074	02	2	
034	43	RCL		075	03	3	
035	16	16		076	71	SBR	Call print
036	85	+		077	85	+	subroutine
037	43	RCL		078	43	RCL	
038	15	15		079	29	29	
039	95	=		080	91	R/S	
040	35	1/X		081	78	Σ+	

Line	Key	Entry	Coments	Line	Key	Entry	Comments
082	91	R/S		123	15	E	
083	76	LBL		124	42	STO	Enter $1/T'$
084	13	C	Calculate E_a	125	29	29	Derive $\ln k'$
085	69	OP		126	01	1	
086	12	12		127	00	0	Call print
087	32	X:T		128	71	SBR	subroutine
088	42	STO		129	85	+	
089	18	18		130	43	RCL	
090	65	×		131	29	29	
091	01	1		132	69	OP	
092	93	.		133	14	14	
093	09	9		134	42	STO	
094	08	8		135	29	29	
095	07	7		136	01	1	
096	95	=		137	01	1	
097	94	+/-		138	71	SBR	
098	42	STO	Call print	139	85	+	
099	19	19	subroutine	140	43	RCL	
100	42	STO		141	29	29	
101	29	29		142	91	R/S	Enter T_1 for
102	02	2		143	76	LBL	Q_{10}
103	04	4		144	10	E'	
104	71	SBR		145	42	STO	Call print
105	85	+		146	29	29	subroutine
106	43	RCL		147	02	2	
107	29	29		148	07	7	
108	91	R/S		149	71	SBR	
109	76	LBL	r	150	85	+	
110	14	D		151	43	RCL	
111	69	OP		152	29	29	
112	13	13		153	85	+	
113	42	STO		154	02	2	
114	29	29	Call print	155	07	7	
115	02	2	subroutine	156	03	3	
116	06	6		157	95	=	
117	71	SBR		158	42	STO	Enter T_2 for
118	85	+		159	20	20	Q_{10}
119	43	RCL		160	91	R/S	
120	29	29		161	42	STO	
121	91	R/S		162	29	29	
122	76	LBL		163	02	2	

Line	Key	Entry	Comments	Line	Key	Entry	Comments
164	08	8		205	00	0	
165	71	SBR		206	07	7	
166	85	+		207	71	SBR	Print
167	43	RCL		208	85	+	subroutine
168	29	29		209	43	RCL	
169	85	+		210	29	29	
170	02	2		211	91	R/S	
171	07	7		212	76	LBL	
172	03	3		213	85	+	
173	95	=		214	87	IFF	
174	42	STO	Calculate	215	08	08	
175	21	21	Q_{10}	216	95	=	
176	65	×		217	43	RCL	
177	43	RCL		218	29	29	
178	20	20		219	92	RTN	
179	95	=		220	76	LBL	
180	42	STO		221	95	=	
181	22	22		222	42	STO	
182	01	1		223	25	25	
183	00	0		224	73	RC*	
184	55	÷		225	25	25	
185	43	RCL		226	69	OP	
186	22	22		227	04	04	
187	65	×		228	43	RCL	
188	93	.		229	29	29	
189	05	5		230	69	OP	
190	00	0		231	06	06	
191	03	3		232	69	OP	
192	65	×		233	00	00	
193	43	RCL		234	98	ADV	
194	19	19		235	92	RTN	
195	65	×					
196	01	1	Q_{10}				
197	00	0					
198	00	0					
199	00	0					
200	95	=					
201	22	INV					
202	23	LNX					
203	42	STO	Call print				
204	29	29	subroutine				

Register Contents, Labels, and Data Cards--Algebraic System

Register	Contents	Labels	Contents
R0	0	Label B'	Pring toggle
R1	$\sum \log k$	Label A	Enter T (°C)
R2	Used	Label B	Enter k
R3	n (counter)	Label C	Calculates E_a
R4	$\sum 1/T$	Label D	Calculates r
R5	$\sum (1/T)^2$	Label E	Enter $1/T$; derive $\ln k'$
R6	$\sum (1/T)(\log k)$		
R9	Used		
R14	10^3		
R15	T (°C)		
R16	273		
R17	k		
R18	Slope		
R19	4.576		

Data Card (4 wnd Write)

Alpha Print Code	Register	Contents
34020100.	07	TEMP
2633765.	10	1/T
27314265.	11	V
37173033.	20	LOGV
2633700.	21	EA
42000000.	22	CORR
27322242.	23	1/T*
17130000.	24	LNV*
15323535.	26	T1
37020000.	27	T2
37030000.	28	Q10

Example

The rate of mitochondrial succinate cytochrome c reductase was assayed as a function of temperature. Calculate the activation energy for the reaction over the temperature span for which the measurements were made.

TABLE 1

Temperature (°C)	(nmoles cyto-chrome c reduced
12	160
15	203
18	243
20	256
22	270
24	298
25	301
28	316

Solution

T (°C)	$1/T \times 10^3$	ln k
12	3.509	5.075
15	3.472	5.313
18	3.436	5.493
20	3.413	5.545
22	3.390	5.600

E_a = 7.115 kcal/mole
r = -0.971
A = 5 × 10^7

Fitted line

$1/T$	$\ln k$
3.5	5.194
3.4	5.552
3.3	5.910

Q_{10} calculation

$$T_1 = 10\,°C$$
$$T_2 = 20\,°C$$
$$Q_{10} = 1.540$$

References

1.   J.M. Lyons, J.K. Raison, and J. Kumamoto (1974), in "Bio-membranes," Part A., *Methods in Enzymology 32,* 258-262.

2.   J.R. Silvius, B.D. Read, and R.N. McElhaney (1978), "Membrane Enzymes: Artifacts in Arrhenius Plots due to Temperature Dependence of Substrate-Binding Affinity," *Science 199,* 902-904.

V
THERMODYNAMICS IN BIOCHEMISTRY

5A. NERNST EQUATION--OXIDATION-REDUCTION POTENTIALS

Redox reactions (oxidation-reduction reactions) play important roles in biochemical systems. Some specific examples of redox reactions are in the electron transport system of mitochondria and the photosynthetic electron transport systems of plants and certain bacterial cells. A redox component (redox couple) consists of a reduced form (reductant), an oxidized form (oxidant), and an appropriate number of electrons (1).

$$X \; reduced \qquad \overset{\rightarrow}{\leftarrow} \qquad X \; oxidized \; + \; ne^- \qquad (1)$$
$$(electron \; donor) \qquad\qquad (electron \; acceptor)$$

An oxidation reaction involves two redox couples which have different affinities for electrons (2). The couple with the higher affinity for electrons will be reduced while its reductant will become oxidized.

$$X \; red_1 \; + \; Y \; ox_2 \qquad \overset{\rightarrow}{\leftarrow} \qquad X \; ox_1 \; + \; Y \; red_2 \qquad (2)$$

A useful way to characterize the tendency of a component to lose an electron(s), is by comparison with a standard reference half cell, which in the case of chemical and biological systems is the hydrogen half cell ($H_2 \rightleftarrows 2H^+ + 2e^-$). The potential difference of such a system can be formulated utilizing the Nernst equation. Equation 3 relates the observed potential E_h to the standard potential E_0 (i.e., the standard redox potential with all components in their standard states (unit activity for solutes at

251

pH 0), plus the natural log of the ratio of concentrations of the
redox components *ox/red*, times the gas constant and the temperature divided by *n* the number of electrons times *F* the Faraday
constant (i.e., the chemical-electrical potential conversion factor).

$$E_h = E_0 + \frac{RT}{nF} \ln \frac{ox}{red} \tag{3}$$

Since most biological reactions do not function at pH 0 and do
not necessarily have unit activities, "standard conditions" must
have different specifications for biological redox systems. The
Nernst equation symbolism is modified such that E_0 becomes E_m
which represents the midpoint potential (i.e., the point of half
reduction where *ox* = *red* and $E_h = E_m$) and the pH is at some defined physiological value around pH 7. In addition, biochemists
use logarithms to the base 10 by convention rather than natural
logarithms (logarithms to the base *e*) format for the Nernst
equation (equation 4).

$$E_h = E_m + 2.303 \frac{RT}{F} \log \frac{ox}{red} \tag{4}$$

Measurements of a system are accomplished experimentally by use
of a potentiometric device consisting of a standard reference
electrode usually a standard calomel electrode (mercury in con-
tact with mercurous chloride paste) which has been calibrated
against a standard hydrogen half cell and a platinum measuring
electrode. The two electrodes are connected to a meter which
gives a direct readout of E_h in millivolts.

The relative redox state of the system is then monitored
simultaneously with the potential measurements by following spectral absorbance changes of the component. Typically, the system
under study is titrated by stepwise additions of reductant followed by reoxidation of the system with an appropriate oxidant.
With the observed *E* and extent of oxidation-reduction values in
hand, a Nernst semilog plot can be constructed and the midpoint
potential can be derived (Figure 5A). The number of electrons
involved in the redox reaction can also be determined from the
slope of the line. If a system consists of several redox components which have similar spectral properties, the Nernst plots
will be nonlinear as in Figure 5A. Such is the case for cytochrome oxidase which has a sigmoidal semilog plot due to the
spectral overlaps of cytochrome *a* and cytochrome a_3. If one
assumes that the spectral contribution of each of the components
is equivalent, then a point halfway between the curves can be
taken to represent 100% oxidation of one component and the

midpoint potentials for the two components can be resolved.
However, if there is more of a contribution of one component than
another, microcomputer analysis is in order (2).

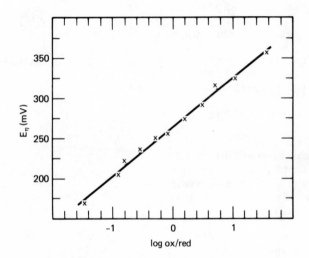

Figure 5A.   A redox plot of cytochrome *c*

User Instructions--RPN

Step	Instructions	Input	Keys	Output
1	Initialize: Clear Registers		f e	0.00
2	To set print flag		f d	1.00
3	To clear print flag		f d	0.00
4	Enter value of 100% Oxidized	100%	f a	100%
5	Enter amount reduced	*red*	ENTER↑	*red*
6	Enter observed potential	*mV*	A	*n*
	To delete incorrect data:			
7	Enter *red*	*red*	ENTER↑	*red*
8	Enter potential	*mV*	B	*n* - 1
	Repeat 5 and 6 for all data			
9	Compute coefficient of determination, number of electrons transferred, and midpoint potential, in millivolts		C	r^2 n E_m
	To compute best fit straight line for plotting:			
10	Enter log (*Ox/red*)	log (*Ox/red*)	D	*mV*

Program Listing--RPN

Line	Key	Comments	Line	Key	Comments
001	*LBLe		046	LOG	
002	CLRG	Initialize: Clear	047	F2?	
003	P≠S	Registers	048	GSB3	
004	CLRG		049	F2?	
005	CLX		050	GSB9	
006	RTN		051	Σ-	
007	*LBL₀		052	RTN	
008	STO0	→ Store 100% Ox	053	*LBLC	
009	5		054	P≠S	
010	9		055	RCL8	
011	.		056	RCL4	Compute regression
012	1		057	RCL6	coefficients
013	STOD	→ Store 2.303 RT/F	058	x	
014	RTN		059	RCL9	
015	*LBLA		060	÷	
016	2		061	-	
017	4		062	ENT↑	
018	5		063	ENT↑	
019	+		064	RCL4	
020	X≠Y	→ E_h	065	X²	
021	ENT↑		066	RCL9	
022	ENT↑		067	÷	
023	RCL0		068	RCL5	
024	X≠Y		069	X≠Y	
025	-		070	-	
026	X≠Y	→ log (Ox/Red)	071	÷	
027	÷		072	STOB	→ slope
028	LOG		073	x	
029	F2?	Print if print	074	RCL6	
030	GSB9	flag is on	075	X²	
031	Σ+	Accumulate sums	076	RCL9	
032	RTN		077	÷	
033	*LBLB	Delete incorrect	078	CHS	
034	2	data	079	RCL7	
035	4		080	+	
036	5		081	÷	
037	+		082	PRTX	→ r^2
038	X≠Y		083	RCL6	
039	ENT↑		084	RCL4	
040	ENT↑		085	RCLB	
041	RCL0		086	x	
042	X≠Y		087	-	
043	-		088	RCL9	
044	X≠Y		089	÷	
045	÷		090	STOA	→ intercept

Line	Key	Comments
091	RCLD	
092	RCLB	
093	÷	
094	PRTX	→ n number of
095	RCLA	electrons
096	PRTX	→ E_m (midpoint)
097	SPC	
098	P≳S	
099	RTN	
100	*LBLD	Projection/plotting
101	RCLB	routine
102	x	
103	RCLA	
104	+	
105	RTN	
106	*LBL9	Print subroutine
107	PRTX	
108	X≳Y	
109	PRTX	
110	SF2	
111	SPC	
112	RTN	
113	*LBL3	Print deletion
114	SPC	indicator
115	DSP0	
116	1	
117	CHS	
118	PRTX	
119	SF2	
120	DSP3	
121	R↓	
122	RTN	
123	*LBLd	Print flag: Set
124	0	and Clear
125	F2?	
126	RTN	
127	1	
128	SF2	
129	RTN	

Register Contents, Labels, and Data Cards--RPN

Register	Contents	Labels	Contents
R_0	100% Ox	A	$Red \rightarrow mV(+)$
R_{s4}	$\sum \log Ox/red$	B	$Red \rightarrow mV(-)$
R_{s5}	$\sum (\log Ox/red)^2$	C	$\rightarrow r^2, n, E_m$
R_{s6}	$\sum mV$	D	$\log (Ox/red) \rightarrow mV$
R_{s7}	$\sum (mV)^2$	a	100% $Ox\uparrow$
R_{s8}	$\sum mV(\log Ox/red)$	d	Print?
R_{s9}	n	e	Initialize
R_A	E_m		
R_B	slope		
R_D	$2.303RT/F$		

User Instructions--Algebraic System

Step	Instructions	Input	Keys	Output
1	Print toggle		2nd C'	
2	Enter value of totally oxidized	100% Ox	2nd A'	
3	Enter value for RT/F		2nd B'	
4	Enter amount reduced	Red	A	$\log Ox/red$
5	Enter log Ox/red into linear regress		R/S	
6	Enter observed milli-volts	mV obs	B	mV

Step	Instructions	Input	Keys	Output
7	Enter mV into linear regress; enter all data pairs before calculation		R/S	
8	Calculate number of electrons passed		C	e
9	Calculate r		D	r
10	Derive fitted line and midpoint (i.e., log Ox/red = 0)	log Ox/red	E	mV'

Program Listing--Algebraic System

Line	Key	Entry	Comments	Line	Key	Entry	Comments
000	76	LBL		026	71	SBR	
001	18	C'	Print	027	85	+	
002	87	IFF	toggle	028	43	RCL	
003	08	08		029	15	15	
004	45	Y^X		030	91	R/S	
005	86	STF		031	76	LBL	
006	08	08		032	17	B'	
007	92	RTN		033	42	STO	RT/F
008	76	LBL		034	20	20	
009	45	Y^X		035	42	STO	
010	22	INV		036	29	29	
011	86	STF		037	02	2	
012	08	08		038	01	1	Call print
013	92	RTN		039	71	SBR	subroutine
014	76	LBL		040	85	+	
015	16	A'	100% Ox	041	91	R/S	
016	42	STO		042	76	LBL	
017	15	15		043	11	A	
018	42	STO		044	42	STO	
019	29	29	Call print	045	16	16	
020	00	0	subroutine	046	42	STO	% red
021	07	7		047	29	29	
022	71	SBR		048	01	1	
023	85	+		049	00	0	
024	00	0		050	71	SBR	Call print
025	08	8		051	85	+	subroutine

Line	Key	Entry	Comments	Line	Key	Entry	Comments
052	53	(		095	05	5	
053	53	(		096	95	=	
054	43	RCL		097	42	STO	
055	15	15		098	29	29	
056	75	-		099	02	2	
057	43	RCL		100	06	6	
058	16	16		101	71	SBR	Call print
059	54	)		102	85	+	subroutine
060	55	÷		103	43	RCL	
061	43	RCL		104	29	29	
062	16	16		105	91	R/S	
063	54	)		106	78	Σ+	Linear
064	28	LOG	log Ox/red	107	92	RTN	regression
065	42	STO		108	76	LBL	$mV\ (E_h)$
066	29	29		109	13	C	
067	02	2		110	69	OP	
068	02	2		111	12	12	
069	71	SBR	Call print	112	32	X:T	
070	85	+	subroutine	113	42	STO	Number of
071	02	2		114	21	21	electrons
072	03	3		115	43	RCL	passed
073	71	SBR		116	20	20	
074	85	+		117	55	÷	
075	43	RCL		118	43	RCL	
076	29	29		119	21	21	
077	91	R/S		120	95	=	
078	36	PGM	Linear	121	42	STO	
079	01	01	regression	122	29	29	
080	32	X:T		123	02	2	
081	92	RTN		124	07	7	
082	76	LBL	Enter mV	125	71	SBR	Call print
083	12	B		126	85	+	subroutine
084	42	STO		127	43	RCL	
085	29	29		128	29	29	
086	02	2		129	91	R/S	
087	04	4		130	76	LBL	r
088	71	SBR	Call print	131	14	D	
089	85	+	subroutine	132	69	OP	
090	43	RCL		133	13	13	
091	29	29		134	91	R/S	
092	85	+	Calomel	135	76	LBL	
093	02	2	electrode	136	15	E	
094	04	4	correction	137	42	STO	

Line	Key	Entry	Comments
138	11	11	
139	69	☐P	log *Ox/red*
140	14	14	→ *mV*
141	42	ST☐	midpoint
142	29	29	
143	02	2	
144	08	8	
145	71	SBR	Call print
146	85	+	subroutine
147	43	RCL	
148	29	29	
149	91	R/S	
150	76	LBL	Print
151	85	+	subroutine
152	87	IFF	
153	08	08	
154	95	=	
155	43	RCL	
156	29	29	
157	92	RTN	
158	76	LBL	
159	95	=	
160	42	ST☐	
161	25	25	
162	73	RC*	
163	25	25	
164	69	☐P	
165	04	04	
166	43	RCL	
167	29	29	
168	69	☐P	
169	06	06	
170	69	☐P	
171	00	00	
172	98	ADV	
173	92	RTN	

Register Contents, Labels, and Data Cards--Algebraic System

Register	Contents	Labels	Contents
R0 → R6	Linear regression	Label A	log *Ox/red*
R11	log *Ox/red*	Label B	*mV*
R15	100% *Ox*	Label C	*e*
R16	*Red*	Label D	*r*
R17	*mV*	Label E	*mV'*
R20	*RT/F*	Label A'	100% *ox*
R21	*e*	Label B'	*RT/F*
		Label C'	Print toggle

Data Card (4 2nd Write)

Alpha Print Code	Register	Contents
2010161.	07	100%
32440000.	08	OX
35171600.	10	RT/F
3537632l.	21	RED
27223244.	22	LGOX
351716.	23	REI'
30423214.	24	MVOB
30420000.	26	MV
54000000.	27	e
30330000.	28	MP

Example

A 30 μM solution of cytochrome c at pH 7.0 was first titrated with sodium dithionite and then reoxidized by reverse titration with ferricyanide using appropriate mediating dyes. Determine the midpoint potential and the number of electrons passed by cytochrome c at 25°C.

TABLE 1

Total oxidation = 154 units (arbitrary units)
2.303 RT/F = 59.2 mV at 25°C.

Amount of reduction (arbitrary units)	Observed mV
4	112
13	80
24	70
38	47
60	30
84	10
100	3
118	-10
133	-24
137	-40
149	-75

Solution

log *Ox/red*	E_h *mV*
1.57	357
1.04	325
0.73	315
0.48	292
0.19	275
-0.08	255
-0.27	248
-0.52	235
-0.80	221
-0.91	205
-1.47	170

number of electrons passed = 0.97
r = 0.998
$E_{m7.0}$ = 264

Fitted line

log *Ox/red*	*mV*
1.5	355
-1.5	172

References

1. P. Dutton (1979), *in* "Biomembranes," Part C: Biological Oxidations, Methods in Enzymology, VII, Academic Press, New York, Chapter 25.

2. T. Ohnishi (1975), *Biochemica   et Biophysica Acta 378*, 475-490.

5B.   DETERMINATION OF ACTIVITY COEFFICIENTS FROM POTENTIOMETRIC DATA

The electromotive force (E.M.F.) of the cell

$$\text{Pt, } H_2 \text{ (1 atm)} \,|\, HCl(m) \,|\, AgCl(s), \text{ Ag}$$

for which the cell reaction is

$$AgCl(s) + \tfrac{1}{2}H_2(g) \rightarrow H^+(aq) + Ag(s) + Cl^-(aq)$$

is given by

$$E = E_0 - \frac{RT}{F} \ln(a_{H^+} \cdot a_{Cl^-})$$

Note that activities a_i are used in preference to concentrations. This equation may be reexpressed in terms of mean ionic activities, which may, in turn, be separated into concentrations m and activity codfficients γ:

$$E = E_0 - \frac{RT}{F} \ln a_\pm^2 = E_0 - \frac{RT}{F}(\ln m_\pm^2 + \ln \gamma_\pm^2)$$

where $m_\pm = \sqrt{m_{H^+} \cdot m_{Cl^-}}$

$\gamma_\pm \quad \sqrt{\gamma_{H^+} \cdot \gamma_{Cl^-}}$

After rearrangement, we obtain:

$$E + \frac{2RT}{F} (\ln m) - E_0 = - \frac{2RT}{F} \ln \gamma_\pm^2 \qquad (1)$$

One form of the Debye-Hückel theory yields an expression for $\log \gamma_\pm$:

$$\log \gamma_\pm = - A\sqrt{m} + C(m)$$

where A and C are constants. Combining this equation with equation 1, inserting numerical values of the constants, and rearranging, we obtain

$$E + 0.1183 \log m - 0.0602\sqrt{m} = E_0 - 0.1183C(m) \qquad (2)$$

Equation 2 can form a graphical basis for the determination of E_0 by plotting the left-hand side as a function of m. The intercept should be E_0. Substitution of E_0 into equation 1 would then yield values of $\gamma_\pm$. Experimentally, this is not straightforward since there is often considerable experimental error owing to the influence of small amounts of impurities. Good experimental data are, however, available in the literature.

For a general electrolyte, we can define mean ionic concentration $m_\pm$ (and, in a similar manner, $\gamma_\pm$) as follows:

$$m_\pm^\nu = m^\nu (\nu_+^{\nu+} \cdot \nu_-^{\nu-})$$

where m = concentration of electrolyte (moles/L)
ν_+ = number of cations
ν_- = number of anions
ν = total number of ions

Alternatively,

$$m_\pm^\nu = fm^\nu$$

where $f = (\nu_+^{\nu+} \cdot \nu_-^{\nu-})$. Thus for $CaCl_2$, $\nu_+ = 1$, $\nu_- = 2$, $\nu = 3$, and $f = 4$.

In the specific case discussed at the beginning of this section, the potential of one of the electrodes was zero since this was the standard hydrogen electrode. In the general case, however, the reference potential may be nonzero, and we will denote it by E_{ref}. Making use of these definitions, we can now express equation 2 in the general form:

$$E + E_{ref} + A(\log m) + B(\log f) - 0.0602\sqrt{m} = E_0 - ACm$$

where $A = 0.059156\nu/n$
n = number of electrons involved in the cell reaction per mole
$B = 0.059156/n$

The calculator program presented here utilizes these relationships. The left-hand side of the equation is taken as the dependent variable, and the molar concentration as the independent variable. Pairs of data entered as molarity and voltage are transformed as appropriate for performing the least squares linear regression calculation (see comments in the listing).

This program is derived from one written by John R. Joyce and published in Hewlett-Packard's *HP-67/97 User's Library Solutions in Chemistry*. This program was based on a Fortran IV program in reference 1. This is an effective illustration of the fact that many problems previously requiring a large computer can now be solved more reasonably on a pocket calculator.

User Instructions--RPN

Step	Instructions	Input	Keys	Output
1	Initialize: Clear Registers		f a	0.00
2	To set print flag		E	1.00
3	To clear print flag		E	0.00
4	Enter f	f	ENTER↑	
5	Enter ν	ν	f b	
6	Enter E_{ref}: E.M.F. of reference electrode	E_{ref}	ENTER↑	
7	Enter n, the number of electrons	n	f c	
8	Enter data: molarity	M	ENTER↑	
	potential	V	A	n
	To delete incorrect data:			
9	Enter: molarity	M	ENTER↑	
	potential	V	B	$n - 1$
10	Repeat step 8 for all data			
11	Compute coefficient of determination and standard electrode potential (E_0)		C	r^2 E_0
12	Compute activity coefficient: Enter: molarity	M	ENTER ↑	
	potential	V	D	γ

Program Listing--RPN

Line	Key	Comments	Line	Key	Comments
001	*LBLa		046	LOG	
002	CLRG	Initialize: Clear	047	x	
003	P≷S	Registers	048	ST+2	
004	CLRG		049	.	
005	CLX		050	0	
006	RTN		051	6	
007	*LBLb		052	0	
008	STOB	Store ν	053	2	
009	R↓		054	RCL1	
010	STOA	Store f	055	√X	
011	RTN		056	x	
012	*LBLc		057	ST-2	
013	STOD	Store number of	058	RCL2	
014	R↓	electrons	059	RCL1	
015	STOC	Store E_{ref}	060	RTN	
016	RTN		061	*LBL2	
017	*LBLA		062	.	
018	F2?	Enter molarity and	063	0	
019	GSB9	potential	064	5	Compute A
020	GSB1		065	9	
021	Σ+	Accumulation of	066	1	
022	RTN	sums after transform	067	5	
023	*LBLB	Data deletion	068	6	
024	F2?		069	RCLB	
025	GSB3	Print deletion indi-	070	x	
026	F2?	cator if print flag	071	RCLD	
027	GSB9	is on	072	÷	
028	GSB1		073	RTN	
029	Σ-	Delete	074	*LBLC	
030	RTN		075	P≷S	
031	*LBL1	Transform entered	076	RCL8	
032	STO2	data into form	077	RCL6	
033	R↓	suitable for linear	078	RCL4	
034	STO1	regression	079	x	
035	RCLC		080	RCL9	Compute; coeffi-
036	ST+2		081	÷	cient of determina-
037	GSB2		082	-	tion and standard
038	RCL1		083	X²	electrode potential,
039	LOG		084	RCL4	E_0
040	x		085	X²	
041	ST+2		086	RCL9	
042	GSB2		087	÷	
043	RCLB		088	CHS	
044	÷		089	RCL5	
045	RCLA		090	+	

Line	Key	Comments	Line	Key	Comments
091	÷		138	LOG	
092	RCL6		139	x	
093	x^2		140	-	
094	RCL9		141	GSB2	
095	÷		142	÷	
096	CHS		143	10^x	
097	RCL7		144	RCL1	
098	+		145	PRTX	→ Molarity
099	÷		146	X⇌Y	
100	PRTX	→ r^2	147	PRTX	Activity coefficient
101	RCL5		148	RTN	→ Y
102	RCL6		149	*LBL9	Print subroutine
103	x		150	X⇌Y	
104	RCL4		151	PRTX	
105	RCL8		152	X⇌Y	
106	x		153	PRTX	
107	-		154	SF2	
108	RCL9		155	SPC	
109	RCL5		156	RTN	
110	x		157	*LBL3	Print deletion indi-
111	RCL4		158	DSP0	cator
112	x^2		159	1	
113	-		160	CHS	
114	÷	Standard electrode	161	PRTX	
115	STOE	potential	162	SF2	
116	PRTX	→ E_0	163	DSP4	
117	SPC		164	R↓	
118	P⇌S		165	RTN	
119	RTN		166	*LBLE	Print flag: Set and
120	*LBLD	Compute:	167	0	clear
121	STO2	activity coeffi-	168	F2?	
122	R↓	cient for entered	169	RTN	
123	STO1	molarity	170	1	
124	RCLE		171	SF2	
125	RCLC		172	RTN	
126	-				
127	RCL2				
128	-				
129	GSB2				
130	RCL1				
131	LOG				
132	x				
133	-				
134	GSB2				
135	RCLB				
136	÷				
137	RCLA				

Y

Register Contents, Labels, and Data Cards--RPN

Register	Contents	Labels	Contents
R_1	Last M	A	$M \uparrow V(+)$
R_2	Last V	B	$M \uparrow V(-)$
R_3	γ	C	$\rightarrow r^2, E_o$
R_{s4}	$\sum M$	D	$M \uparrow V \rightarrow \gamma$
R_{s5}	$\sum M^2$	E	Print?
R_{s6}	$\sum y$	a	Initialize
R_{s7}	$\sum y^2$	b	$f \uparrow \nu$
R_{s8}	$\sum My$	c	$E_{ref} \uparrow$ number of e^-
R_{s9}	n		
R_A	f		
R_B	ν		
R_C	E_{ref}		
R_D	Number of e^-		
R_E	E_0		

User Instructions--Algebraic System

Step	Instructions	Input	Keys	Output
1	Enter number of e^-	e^-	2nd A'	
2	Enter ν	ν	2nd B'	
3	Enter E_{ref}	E_{ref}	2nd C'	
4	Enter f	f	2nd D'	B ($\log f$)
5	Enter M	M	A	
6	Enter V	V	B	n
7	Compute E_0		C	E_0
8	Compute correlation coefficient		R/S	r
9	Enter V		D	
10	Enter M; calculate activity coefficient	M		γ

Program Listing--Algebraic System

Line	Key	Entry	Comments	Line	Key	Entry	Comments
000	76	LBL		015	05	5	
001	16	A'	Enter e^-	016	09	9	
002	42	STO		017	01	1	
003	10	10		018	05	5	
004	91	R/S	Enter ν	019	06	6	
005	76	LBL		020	95	=	A
006	17	B'		021	42	STO	
007	42	STO		022	12	12	
008	11	11		023	93	.	
009	55	÷		024	00	0	
010	43	RCL		025	05	5	
011	10	10		026	09	9	
012	65	×		027	01	1	
013	93	.		028	05	5	
014	00	0		029	06	6	

Line	Key	Entry	Comments	Line	Key	Entry	Comments
030	55	÷		071	42	STO	
031	43	RCL		072	16	16	
032	10	10		073	43	RCL	linear
033	95	=		074	19	19	regression
034	42	STO		075	36	PGM	
035	13	13		076	01	01	
036	91	R/S	*B*	077	32	X:T	
037	76	LBL		078	91	R/S	
038	18	C'		079	76	LBL	*E*
039	42	STO	Enter E_{ref}	080	12	B	
040	14	14		081	42	STO	
041	91	R/S		082	17	17	
042	76	LBL		083	85	+	
043	19	D'	Enter *f*	084	43	RCL	
044	42	STO		085	14	14	
045	15	15		086	85	+	
046	28	LOG		087	53	(	
047	65	×		088	43	RCL	
048	43	RCL		089	12	12	
049	13	13		090	65	×	
050	95	=		091	53	(	
051	42	STO	*B* (log *f*)	092	43	RCL	
052	13	13		093	19	19	
053	91	R/S		094	28	LOG	
054	76	LBL		095	54	)	
055	10	E'		096	54	)	
056	42	STO		097	85	+	
057	14	14		098	43	RCL	
058	91	R/S		099	13	13	
059	76	LBL		100	75	-	
060	11	A	Enter *M*	101	43	RCL	
061	42	STO		102	16	16	
062	19	19		103	95	=	
063	34	ГX		104	42	STO	
064	65	×		105	18	18	
065	93	.		106	78	Σ+	
066	00	0		107	91	R/S	
067	06	6		108	76	LBL	Standard
068	00	0		109	13	C	electrode
069	02	2		110	69	OP	potential
070	95	=		111	12	12	

Line	Key	Entry	Comments	Line	Key	Entry	Comments
112	42	STO		154	28	LOG	
113	16	16		155	95	=	
114	91	R/S	$\rightarrow E_0$	156	55	÷	
115	69	OP		157	43	RCL	
116	13	13	$\rightarrow r^2$	158	12	12	
117	91	R/S		159	95	=	
118	76	LBL		160	22	INV	
119	14	D		161	28	LOG	
120	42	STO	Compute acti-	162	91	R/S	
121	07	07	vity coeffic-				
122	91	R/S	ient entered				
123	42	STO	molarity				
124	08	08					
125	53	(					
126	43	RCL					
127	16	16					
128	75	-					
129	43	RCL					
130	14	14					
131	75	-					
132	43	RCL					
133	08	08					
134	54	)					
135	75	-					
136	53	(					
137	43	RCL					
138	12	12					
139	65	×					
140	43	RCL					
141	07	07					
142	28	LOG					
143	54	)					
144	75	-					
145	53	(					
146	43	RCL					
147	12	12					
148	55	÷					
149	43	RCL					
150	11	11					
151	65	×					
152	43	RCL	$\rightarrow \gamma$				
153	15	15					

Register Contents, Labels, and Data Cards--Algebraic System

Register	Contents	Labels	Contents
R0 → R6	Linear regression	Label A'	Enter number of e^-
R10	number of e^-	Label B'	Enter ν
R11	ν	Label C'	Enter E_{ref}
R12	A	Label D'	Enter f
R13	B (log f)		
R14	E_{ref}	Label A	Enter M
R15	M	Label B	Enter V
R16	$0.0602\sqrt{M}$	Label C	Calculate E_0
R17	V	Label D	Enter V
R18		Label E	Enter M
R19	E_0		Calculate γ
R20	V		
R21	M		

Example

The data presented here for analysis were obtained for the cell reaction mentioned in the beginning of this section:

$$AgCl(s) + \tfrac{1}{2}H_2(g) \rightarrow H^+(aq) + Ag(s) + Cl^-(aq)$$

In this case $f = 1$, $\nu = 2$, $n = 1$, and $E_{ref} = 0.00$ since this is the standard hydrogen electrode. Calculate the standard electrode potential, E_0 and the activity coefficients at the indicated concentrations

TABLE 1

Concentration (M)	Potential (V)
0.0045	0.5038
0.0056	0.4926
0.0073	0.4795
0.0091	0.4686
0.0112	0.4586
0.0134	0.4497
0.0171	0.4378
0.0256	0.4182
0.0539	0.3822
0.1238	0.3420

$r^2 = 0.9974$
$E_0 = 0.2221$

Solution

Concentration (M)	Activity Coefficient, γ
0.0045	0.9246
0.0056	0.9240
0.0073	0.9146
0.0091	0.9071
0.0112	0.8954
0.0134	0.8899
0.0171	0.8791
0.0256	0.8599
0.0539	0.8230
0.1238	0.7835

References

1. G. Beech (1975), *FORTRAN IV in Chemistry,* John Wiley & Sons, Inc., New York, pp. 64-66.

2. J.R. Joyce (1977), "Activity Coefficients from Potentiometric Data," in Hewlett-Packard's *HP-67/97 User's Library Solutions in Chemistry.*

VI

SPECTROSCOPY

6A. BEER-LAMBERT LAW

A close relation exists between the extent of absorption and the concentration of the absorbing substance. The change in the intensity of light with distance through the solution can be written as

$$\frac{dI}{dx} = -KCx \tag{1}$$

where C is the concentration of the absorbing substance in solution, I is the intensity of the light at distance x, and K is a constant characteristic of the wavelength of the light and of the substance. The negative sign indicates that the intensity of the light decreases with increasing distance.

Rearranging equation 1 and integrating between the limits of x equal to zero, where I_0 is the intensity of the incident light, I the intensity of light after passing through a transparent solution, and x the distance the light travels through the transparent solution, gives

$$\ln \frac{I}{I_0} = -KCx \tag{2}$$

Equation 2 can be expressed in logarithms to the base 10, and

$$\log \frac{I_0}{I} = \varepsilon C d = \text{optical density} \qquad (3)$$

The quantity $\log(I_0/I)$ is known as optical density or absorbance, A; ε is the molar extinction coefficient with units of liters per mole per centimeter.

Equation 3 is an expression of the Beer-Lambert law. The percent transmission ($\%T$) is defined as $100 I/I_0$, and the relation between percent transmission and optical density is

$$A = \log \frac{100}{\%T} = 2 - \log \%T \qquad (4)$$

The two coefficients which are sufficient to describe the interaction of radiation with a medium are the refractive index and the absorption coefficient.

According to equation 3, the plot of absorbance against concentration should yield a straight line; it frequently does, but not always. If the light filter or monochromator of the spectrophotometer is not sufficiently discriminating so that the light used covers a wide band of wavelengths, deviation from Beer's law can occur. At low concentration of solute the decrease in the light transmitted is a function primarily of the peak absorption with a characteristic extinction coefficient. At high concentrations, however, Beer's law is often violated because of scattering or structural changes (e.g., dimerization, aggregation, or chemical reaction). Figure 6A shows positive and negative deviation from Beer's law and the causes.

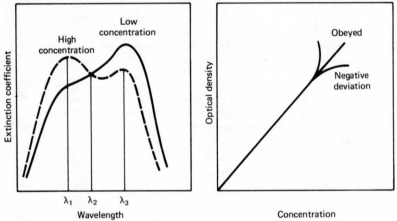

Figure 6A.   Beer-Lambert Law--positive and negative deviation.

At the left is a spectral shift associated with increasing con-
centration--often a result of polymerization. Note that at one
wavelength, λ_2, there is no change in molar extinction coefficient with change in concentration. This wavelength is called
the *isosbestic point*. At the right is a curve showing deviation
from Beer's law. At λ_1, the deviation is positive, and at λ_3 it
is negative. At the isosbestic point, λ_2, Beer's law is always
obeyed.

The program presented here allows the user to interchange-
ably solve for either concentration or absorbance if the extinction coefficient is known. The distance, or pathlength, of the
cell is assumed to be 1 cm. The user may alter the pathlength by
entering the new value of d with the f a user-definable key.

If the extinction coefficient is unknown, it can be computed
by entering a series of absorbance or $\%T$ values at different concentrations. The data are entered into the linear regression
routine, which allows the computation of the best fit extinction
coefficient, the coefficient of determination, and an error factor, i, which reflects the fact that the data do not pass per-
fectly through the origin. Also, the program allows for the
direct interconversion of absorbance and $\%T$ values.

User Instructions--RPN

Step	Instructions	Input	Keys	Output
1	Load program			0.00
2	Option 1: Initialize default pathlength is 1 cm		f a	1.0
3	Option 2: If pathlength is not 1 cm, enter new path-length	d	f a	d
4	Option 3: If mM extinction coefficient is known and pathlength is not 1 cm, enter ε and new pathlength	ε ↑ d	f a	d
5	Option 4: If extinction coefficient entered is expressed as $E_{1\%}$ [i.e., absorbance of 1%(w/v) solution of protein]		f b	1.00
6	If absorbance is known, Enter A	A	A	A
7	Compute concentration; express concentration as mM if $E_{1\%}$ flag is not set		B	
8	If concentration is known, enter C	C	B	C
9	Compute absorbance		A	A
10	For data where absorbance and concentration are known: enter C_i and A_i	C_i A_i	ENTER↑ C	n
11	To delete incorrect data pair: enter incorrect C_i and A_i	C_i A_i	ENTER↑ f c	$n - 1$

Step	Instructions	Input	Keys	Output
12	After all data are entered: compute ε, i, r		R/S	ε i r
	Press R/S again to display ε, i, r again			
13	If data are in the form of concentration and percent transmittance: enter C_i and $\%T_i$	C_i $\%T_i$	ENTER D	n
14	To delete incorrect data pair: enter incorrect C_i and $\%T_i$	C_i $\%T_i$	ENTER f d	$n - 1$
15	After all data are entered: compute ε, i, r		R/S	ε i r
16	To convert absorbance to $\%T$	A	E	$\%T$
17	To convert $\%T$ to absorbance	$\%T$	f e	A

Program Listing--RPN

Line	Key	Comments	Line	Key	Comments
001	*LBLa		018	*LBLA	
002	CLRG	Initialize	019	DSP3	
003	P⇄S		020	STO0	
004	CLRG		021	F3?	
005	F3?	d entered?	022	R/S	
006	GTO1	If not, default	023	*LBL2	Absorbance input?
007	GTO0	$d = 1$ cm	024	F0?	If not, compute absor-
008	*LBL1		025	GSB5	bance and store in R_0
009	STOA	Store 1 in R_A	026	RCLA	
010	R↓		027	RCLB	
011	STOB	Store ε in R_B	028	RCL1	
012	RCLA	(optional)	029	x	
013	RTN		030	x	
014	*LBL0	Default $d = 1$ cm	031	STO0	
015	1		032	R/S	
016	STOA		033	*LBLB	
017	RTN		034	STO1	

Line	Key	Comments	Line	Key	Comments
035	F3?	Concentration in-	081	Σ+	Accumulate sums
036	R/S	put?	082	FIX	
037	*LBL3	If not, compute	083	DSP4	
038	RCL0	concentration and	084	R/S	
039	RCLA	store in R_1	085	GT07	
040	RCLB		086	*LBLd	
041	x		087	GSBe	Convert $\%T \to A$
042	÷		088	Σ-	Deletion of data
043	F0?		089	FIX	pair
044	GSB6		090	DSP4	
045	ST01		091	R/S	
046	RTN		092	GT07	
047	*LBLb	Set extinction	093	*LBL7	
048	F0?	coefficient as $E_{1\%}$	094	P≷S	Compute and display
049	GT04		095	RCL9	r, ε, i
050	SF0	Flag 0 on = 1.00	096	RCL8	
051	1	Flag 0 off = 0.00	097	x	
052	RTN	ε is assumed to be	098	RCL4	
053	*LBL4	in units of liters/	099	RCL6	
054	CF0	mole·sec	100	x	
055	0		101	-	
056	RTN		102	STOC	
057	*LBL5		103	RCL9	
058	1		104	RCL7	
059	0		105	x	
060	ST÷1		106	RCL6	
061	RTN		107	X²	
062	*LBL6		108	-	
063	1		109	STOD	
064	0		110	÷	
065	x		111	RCLA	
066	RTN		112	÷	
067	*LBLC	Accumulate sums for	113	ENG	
068	Σ+	linear regression	114	PRTX	Display ε
069	FIX		115	RCL7	
070	DSP4		116	RCL4	
071	R/S		117	x	
072	GT07		118	RCL6	
073	*LBLc		119	RCL8	
074	Σ-	Deletion of data	120	x	
075	FIX	pair	121	-	
076	DSP4		122	RCLD	
077	R/S		123	÷	
078	GT07		124	FIX	
079	*LBLD		125	PRTX	Display i
080	GSBe	Convert $\%T \to A$	126	RCLC	

Line	Key	Comments
127	RCLD	
128	√X	
129	÷	
130	RCL9	
131	RCL5	
132	x	
133	RCL4	
134	X²	
135	-	
136	√X	
137	÷	
138	P≷S	
139	R/S	Display r
140	GTO7	
141	RTN	
142	*LBLE	
143	2	$A \to \%T$
144	-	
145	CHS	
146	10ˣ	
147	FIX	
148	DSP1	
149	RTN	
150	*LBLe	$\%T \to A$
151	LOG	
152	2	
153	-	
154	CHS	
155	FIX	
156	DSP4	
157	RTN	

Register Contents, Labels, Data Cards--PRN

Register	Contents	Labels	Contents
R_0	A	A	$\leftrightarrow A$
R_1	C	B	$\leftrightarrow C$
R_{s4}	$\sum C$	C	$C \uparrow A(+)$
R_{s5}	$\sum C^2$	D	$C \uparrow T(+)$
R_{s6}	$\sum A$	E	$A \rightarrow \%T$
R_{s7}	$\sum A^2$	a	$\varepsilon \uparrow d \rightarrow \text{init}$
R_{s8}	$\sum CA$	b	$E_{1\%}$
R_{s9}	n	c	$C \uparrow A(-)$
R_A	d	d	$C \uparrow T(-)$
R_B	ε	e	$\%T \rightarrow A$

User Instructions--Algebraic System

Step	Instructions	Input	Keys	Output
1	Option 1: Enter path-length, initialize	path-length	2nd A'	
2	Option 2: Enter mM extinction coefficient if known	ε	2nd B'	
3	Option 3: Enter $E_{1\%}$ extinction coefficient	$E_{1\%}$	2nd C'	
4	If absorbance is known, enter A and compute concentration	A	A	C
5	If concentration is known, enter C and compute A	C	B	A
6	Enter A for linear regression	A	C	
7	Enter C for linear regression; enter all data pairs before ε, i, r calculation	C	R/S	n
8	Calculate error term i		R/S	i
9	Calculate ε		R/S	ε
10	Calculate correlation coefficient, r		R/S	r
11	Optional entry of %T for linear regression	%T	D	
12	Convert A to %T	A	E	%T
13	Convert %T to A	%T	2nd E'	A

Program Listing--Algebraic System

Line	Key	Entry	Comments	Line	Key	Entry	Comments
000	76	LBL		041	55	÷	
001	16	A'	Enter path-	042	43	RCL	Compute C
002	67	EQ	length or	043	12	12	(liters/mole·
003	38	SIN	default to	044	95	=	cm)
004	42	STO	1 cm path-	045	87	IFF	
005	10	10	length	046	01	01	Flag for 1%
006	91	R/S		047	39	COS	conversion to
007	76	LBL		048	91	R/S	mg/ml
008	38	SIN		049	76	LBL	
009	01	1		050	39	COS	
010	42	STO		051	65	×	
011	10	10		052	01	1	Compute C
012	91	R/S		053	00	0	(mg/ml)
013	76	LBL	Enter ε	054	95	=	
014	17	B'		055	91	R/S	
015	42	STO		056	76	LBL	Enter C;
016	12	12		057	12	B	compute ab-
017	65	×		058	42	STO	sorbance
018	43	RCL		059	15	15	
019	10	10		060	65	×	
020	95	=	ε × path-	061	43	RCL	
021	42	STO	length	062	12	12	
022	12	12		063	95	=	
023	91	R/S		064	42	STO	
024	76	LBL		065	16	16	
025	18	C'		066	91	R/S	
026	42	STO	$E_{1\%}$	067	76	LBL	Enter absorb-
027	12	12		068	13	C	bance for
028	65	×		069	42	STO	linear regres-
029	43	RCL		070	16	16	sion
030	10	10		071	91	R/S	
031	95	=		072	42	STO	Calculate
032	42	STO	ε × path-	073	17	17	linear regres-
033	12	12	length	074	36	PGM	sion
034	86	STF		075	01	01	
035	01	01	Flag for $E_{1\%}$	076	32	X:T	Enter C for
036	91	R/S		077	92	RTN	linear regres-
037	76	LBL		078	43	RCL	sion
038	11	A	Enter absor-	079	16	16	
039	42	STO	bance	080	78	Σ+	
040	13	13		081	92	RTN	

Line	Key	Entry	Comments	Line	Key	Entry	Comments
082	91	R/S		123	42	STO	
083	69	OP		124	19	19	
084	12	12	i	125	02	2	
085	91	R/S		126	75	-	
086	32	X:T		127	43	RCL	
087	55	÷		128	19	19	
088	43	RCL		129	95	=	
089	10	10		130	91	R/S	
090	95	=	ε				
091	91	R/S					
092	69	OP					
093	13	13	r				
094	91	R/S					
095	76	LBL	Enter %T for				
096	14	D	linear re-				
097	28	LOG	gression				
098	42	STO					
099	17	17					
100	02	2					
101	75	-					
102	43	RCL					
103	17	17					
104	95	=					
105	61	GTO					
106	13	C					
107	91	R/S					
108	76	LBL					
109	15	E	A to %T				
110	42	STO					
111	18	18					
112	02	2					
113	75	-					
114	43	RCL					
115	18	18					
116	95	=					
117	22	INV					
118	28	LOG					
119	91	R/S	%T to A				
120	76	LBL					
121	10	E'					
122	28	LOG					

Register Contents, Labels, and Data Cards--Algebraic System

Register	Contents	Labels	Contents
R0 → R6	Linear regression	Label A'	Enter pathlength
R10	Pathlength	Label B'	Enter ε
R11	ε	Label C'	Enter $E_{1\%}$
R12	$E_{1\%}$	Label E'	Enter %T
R13	A	Label A	Enter A for linear regression
R15	C	Label B	Enter C for linear regression
R16	A		
		Label C	Enter A
R17	%T		
		Label D	Enter %T
R18	A		
		Label E	Enter A
R19	%T		

Example

Below are the data obtained from a measurement of inorganic phosphate, P_i by the "ascorbic acid" method at 880 nm. The light path, $d = 1.2$ cm. Compute the extinction coefficient for P_i at 880 nm

Table 1

%T	A	$[P_i]$ (M)
97.9	0.0092	0
58.0	0.2366	8.07×10^{-6}
37.2	0.4295	16.14×10^{-6}
23.1	0.6364	24.21×10^{-6}
14.5	0.8386	32.28×10^{-6}
9.0	1.0458	40.35×10^{-6}

Solution

$$\varepsilon = 21.230 \times 10^3$$
$$i = 0.0187$$
$$r = 0.9998$$

References

1. Henry B. Bull (1964), *An Introduction to Physical Biochemistry*, F.A. Davis Co., Philadelphia, Pa., pp. 192-193.

2. I.H. Segel (1976), *Biochemical Calculations*, 2nd Ed., John Wiley & Sons, Inc., New York, pp. 324-329.

3. David Freifelder (1976), *Physical Biochemistry--Applications to Biochemistry and Molecular Biology*, W.H. Freeman and Company, San Francisco, pp. 380-383.

6B. MULTICOMPONENT SPECTROSCOPY

This method of analysis can be applied to ultraviolet, visible, or infrared spectroscopy. If two absorbing solutes are present in a solution, their concentrations may be determined by solving two simultaneous equations. Similarly, the concentrations of n absorbing solutes can be determined by solving n simultaneous equations, subject to the following conditions:

1. Each component obeys the Beer-Lambert law.
2. The wavelengths of maximum absorbance are sufficiently different for the two substances.

The method is most easily illustrated for a system of two solutes. The absorbance at wavelengths 1 and 2 for substances A and B are given by

$$O.D._1 = \epsilon_1^A \, bC_A + \epsilon_1^B \, bC_B$$

$$O.D._2 = \epsilon_2^A \, bC_A + \epsilon_2^B \, bC_B$$

where ϵ_1^A and ϵ_1^B are the molar extinction coefficients of substances A and B at wavelength 1, ϵ_2^A and ϵ_2^B are the molar extinction coefficients of substances A and B at wavelength 2, b is the pathlength and is assumed to be 1 cm, and C_A and C_B are the concentrations of substances A and B, respectively. If we can measure the optical densities at wavelengths 1 and 2, and if we know or can calculate the molar extinction coefficients, then the concentrations can be calculated.

The calculator program presented here solves two simultaneous equations after the required data are entered.

User Instructions--RPN

Step	Instructions	Input	Keys	Output
1	Enter O.D. at λ_1	$O.D._1$	ENTER↑	
2	Enter ε of A at λ_1	ε_1^A	ENTER↑	
3	Enter ε of B at λ_1	ε_1^B	A	
4	Enter O.D. at λ_2	$O.D._2$	ENTER↑	
5	Enter ε of A at λ_2	ε_2^A	ENTER↑	
6	Enter ε of B at λ_2	ε_2^B	B	
7	Calculate concentration of A		C	c_A
8	Calculate concentration of B		D	c_B

Program Listing--RPN

Line	Key	Comments		Line	Key	Comments
001	*LBLA			045	÷	
002	STO2	Store ε_1^B		046	RTN	
003	R↓					
004	STO1	Store ε_1^A				
005	R↓					
006	STO0	Store O.D.$_1$				
007	RTN					
008	*LBLB					
009	STO5	Store ε_2^B				
010	R↓					
011	STO4	Store ε_2^A				
012	R↓					
013	STO3	Store O.D. $_2$				
014	RTN					
015	*LBLC					
016	RCL1					
017	RCL5	Compute concentra-				
018	x	tion of A				
019	RCL2					
020	RCL4					
021	x					
022	-					
023	STO6					
024	RCL0					
025	RCL5					
026	x					
027	RCL2					
028	RCL3					
029	x					
030	-					
031	RCL6					
032	÷					
033	STOA					
034	ENG					
035	RTN					
036	*LBLD					
037	RCL3					
038	RCL1					
039	x	Compute concentra-				
040	RCL0	tion of B				
041	RCL4					
042	x					
043	-					
044	RCL6					

Register Contents, Labels, and Data Cards--RPN

Register	Contents	Labels	Contents
R_0	$O.D._1$	A	$O.D._1 \uparrow \varepsilon_1^A \uparrow \varepsilon_1^B$
R_1	ε_1^A	B	$O.D._2 \uparrow \varepsilon_2^A \uparrow \varepsilon_2^B$
R_2	ε_1^B	C	$\rightarrow C_A$
R_3	$O.D._2$	D	$\rightarrow C_B$
R_4	ε_2^A		
R_5	ε_2^B		
R_6	Determinant		
R_A	C_A		
R_B	C_B		

User Instructions--Algebraic System

Step	Instructions	Input	Keys	Output
1	Enter O.D. at wavelength 1	$O.D._1$	A	
2	Enter extinction coefficient of A at wavelength 1	ε_1^A	R/S	
3	Enter molar extinction coefficient of B at wavelength 1	ε_1^B	R/S	
4	Enter O.D. at wavelength 2	$O.D._2$	B	
5	Enter molar extinction coefficient of A at wavelength 2	ε_2^A	R/S	
6	Enter molar extinction coefficient of B at wavelength 2	ε_2^B	R/S	
	Calculate concentration of A		R/S	C_A
7	Calculate concentration of B		R/S	C_B

Program Listing--Algebraic System

Line	Key	Entry	Comments	Line	Key	Entry	Comments
000	76	LBL		041	00	00	
001	11	A	Enter O.D.$_1$	042	65	×	
002	42	STD		043	43	RCL	
003	00	00		044	05	05	
004	91	R/S		045	54	)	
005	42	STD	Enter ε_1^A	046	75	-	
006	01	01		047	53	(	
007	91	R/S		048	43	RCL	
008	42	STD		049	02	02	
009	02	02	Enter ε_1^B	050	65	×	
010	91	R/S		051	43	RCL	
011	76	LBL		052	03	03	
012	12	B		053	54	)	
013	42	STD	Enter O.D.$_2$	054	95	=	
014	03	03		055	42	STD	
015	91	R/S		056	13	13	
016	42	STD	Enter ε_2^A	057	43	RCL	
017	04	04		058	13	13	
018	91	R/S		059	55	÷	
019	42	STD	Enter ε_2^B	060	43	RCL	
020	05	05		061	06	06	Compute and
021	53	(		062	95	=	display con-
022	43	RCL	Calculate	063	57	ENG	centration of
023	01	01	determinant	064	42	STD	A
024	65	×	of matrix	065	10	10	
025	43	RCL		066	91	R/S	
026	05	05		067	53	(	
027	54	)		068	43	RCL	
028	75	-		069	03	03	
029	53	(		070	65	×	
030	43	RCL		071	43	RCL	
031	02	02		072	01	01	
032	65	×		073	54	)	
033	43	RCL		074	75	-	
034	04	04		075	53	(	
035	54	)		076	43	RCL	
036	95	=		077	00	00	
037	42	STD		078	65	×	
038	06	06		079	43	RCL	
039	53	(		080	04	04	
040	43	RCL		081	54	)	

Line	Key	Entry	Comments
082	95	=	
083	42	STO	
084	12	12	
085	43	RCL	
086	12	12	
087	55	÷	
088	43	RCL	
089	06	06	
090	95	=	
091	57	ENG	Compute and
092	42	STO	display con-
093	11	11	centration of
094	91	R/S	B

Register Contents, Labels, and Data Cards--Algebraic System

Register	Contents	Labels	Contents
R0	$O.D._1$	Label A	Enter $O.D._1$
R1	ε_1^A	Label B	Enter $O.D._2$
R2	ε_1^B		
R3	$O.D._2$		
R4	ε_2^A		
R5	ε_2^B		
R10	C_A		
R11	C_B		

Example

A. A solution containing two substances, A and B, has an absorbance in a 1 cm cuvette of 0.42 at 450 nm and 0.225 at 260 nm. The molar absorption coefficients of A and B at the two wavelengths are given below. Calculate the concentrations of A and B in the solution.

Compound	ε	
	450 nm	260 nm
A	11,300	37,000
B	15,400	3,000

Solution

$$c_A = 4.115 \times 10^{-6} \, M$$
$$c_B = 24.25 \times 10^{-6} \, M$$

B. The molar extinction coefficients of substance C at 260 and 280 nm are 5248 and 3150, respectively. In isolating C, a reagent D is used whose molar extinction coefficients at 260 and 280 nm are 311 and 350. After isolating A, O.D.$_{260}$ = 2.50 and O.D.$_{280}$ = 2.00. What are the concentrations of C and D?

Solution

$$c_C = 295.2 \times 10^{-6} \, M$$
$$c_D = 3.058 \times 10^{-6} \, M$$

References

1. Henry B. Bull (1964), *An Introduction to Physical Biochemistry,* F.A. Davis Company, Philadelphia, Pa., pp. 192-193.

2. I.H. Segel (1976), *Biochemical Calculations,* 2nd Ed., John Wiley & Sons, Inc., New York, pp. 324-329.

3. David Freifelder (1976), *Physical Biochemistry--Applications to Biochemistry and Molecular Biology,* W.H. Freeman and Company, San Francisco, pp. 380-383.

4. G. Beech (1975), *FORTRAN IV in Chemistry,* John Wiley & Sons, Inc., New York, pp. 75-80.

6C. RADIATIVE FLUORESCENCE LIFETIME

A variety of formulas have been derived to relate the radiative lifetime τ_r to the extinction coefficient for absorption. The simplest to use is the one given by Bowen and Wokes (3). Translated into units of reciprocal micrometers (μm^{-1}), this takes the form

$$\frac{1}{\tau_r} = 2900n^2\bar{\nu}_0^2 \int_0 \varepsilon \, d\bar{\nu} \qquad (\tau_r \text{ in seconds}) \qquad (1)$$

where $\int \varepsilon \, d\bar{\nu}$ is the area under the curve of molecular extinction coefficient plotted against wavenumber, $\bar{\nu}_0$ is the wavenumber of the maximum absorption band, and n is the refractive index of the solvent. For many purposes this equation gives a sufficiently accurate value of τ_r.

It should be noted that the lifetimes calculated in this way are the *radiative* lifetimes, that is, the lifetimes that would be observed in the absence of all other processes by which the molecule can return to the ground state. The observed lifetimes are nearly always less than the calculated values because of competing radiationless processes. The radiative lifetime of the molecule in its lowest excited singlet state is one of the main factors governing the fluorescence intensity observed in solution at room temperature. If the lifetime is exceptionally long, there is a much greater chance of the radiationless processes competing successfully with the radiative process that gives rise to fluorescence, and the intensity of the latter will therefore be lower.

In applying equation 1, the calculator program performs an integration using the trapezoidal rule. It must be remembered that the integration has to be carried out only over the first absorption band.

User Instructions--RPN

Step	Instructions	Input	Keys	Output
1	Initialize: Clear Registers		f e	0.00
2	Enter wavelength of maximum absorption	λ_{max}	f a	λ_{max}
3	Enter step size for trapezoidal integration	$\Delta\lambda$	A	$\Delta\lambda$
4	Enter O.D. at intervals of $\Delta\lambda$ over the first absorption band	$O.D._i$	B	n
5	Calculate approximate radiative lifetime, τ_r		C	τ_r

Program Listing--RPN

Line	Key	Comments	Line	Key	Comments
01♦LBL ε		Initialize: Clear	43♦LBL d		
02 7CLREG		registers and store	44 RCL 23		
03 7P<>S		proportionality	45 *		
04 7CLREG		factor	46 X<>Y		
05 1.9523 E4-			47 /		
06 STO 01			48 RCL 01		
07 CLX			49 /		
08 7FIX			50 STO 04		
09 7DSP2			51 RCL 02		
10 RTN			52 E3-		
11♦LBL a			53 *		Compute μm^{-1} for
12 STO 02		Store λ_{max}	54 1/X		λ_{max}
13 CLX			55 X↑2		
14 RTN			56 1.33		
15♦LBL A			57 X↑2		
16 STO 23		Store step size	58 *		
17 RTN			59 2900		
18♦LBL B			60 *		
19 STO 00		Input $O.D._0$	61 RCL 04		
20 0			62 *		
21 STO 03		Initialize counter	63 1/X		
22♦LBL 09			64 7SCI		
23 RTN			65 7DSP3		
24♦LBL B			66 7PRTX		Compute and display
25 STO 20		Input $O.D._j$, j odd	67 RTN		τ_r
26 XEQ 06		$R_0 \leftarrow R_0 + 2O.D._i$	68♦LBL 06		Subroutine for
27 1			69 ENTER↑		trapezoidal sum-
28 ST+ 03		$n \leftarrow n + 1$	70 +		mation
29 RCL 03			71 ST+ 00		
30 RTN			72 RTN		
31♦LBL B			73 STOP		
32 STO 20		Input $O.D._j$, j even			
33 XEQ 06		$R_0 \leftarrow R_0 + 2O.D._i$			
34 1					
35 ST+ 03		$n \leftarrow n + 1$			
36 RCL 03					
37 GTO 09		Exit routine			
38♦LBL C					
39 2		Compute trapezoidal			
40 RCL 00		area			
41 RCL 20					
42 -					

Register Contents, Labels, and Data Cards--RPN

Register	Contents	Labels	Contents
R_0	Used	A	$\Delta\lambda$ ↑
R_1	1.9523×10^{-4}	B	$O.D._j$ ↑
R_2	λ_{max}	C	→ τ_r
R_3	n	a	λ_{max} ↑
R_A	$O.D._j$	e	Initialize
R_D	$\Delta\lambda$		

User Instructions--Algebraic System

Step	Instructions	Input	Keys	Output
1	Enter maximum wavelength	λ_{max}	2nd A'	
2	Enter refractive index of medium	n	2nd B'	
3	Enter number of sub-intervals	h	R/S	h
4	Enter integration step size	$\Delta\lambda$	R/S	$\Delta\lambda$
5	Enter absorbance at interval of $\Delta\lambda$	$O.D._i$	R/S	$O.D._i$
6	Calculate approximate radiative lifetime		E	τ_r

Program Listing--Algebraic System

Line	Key	Entry	Comments	Line	Key	Entry	Comments
000	76	LBL		041	76	LBL	
001	16	A'	Enter λ_{max}	042	15	E	
002	42	STO		043	36	PGM	Simpson's
003	25	25		044	10	10	rule sub-
004	01	1		045	14	D	routine
005	00	0		046	42	STO	
006	00	0		047	13	13	
007	00	0		048	55	÷	
008	55	÷		049	01	1	
009	43	RCL		050	93	.	
010	25	25		051	09	9	
011	95	=		052	05	5	
012	33	X²	$\bar{\nu}_0^2$	053	02	2	
013	42	STO		054	03	3	
014	26	26		055	52	EE	
015	91	R/S		056	04	4	
016	76	LBL		057	94	+/-	
017	17	B'	Enter refrac-	058	95	=	
018	33	X²	tive index, n	059	42	STO	
019	42	STO		060	28	28	
020	27	27		061	65	×	
021	91	R/S	Enter number	062	43	RCL	
022	36	PGM	of subinter-	063	26	26	
023	10	10	vals	064	65	×	
024	11	A		065	02	2	
025	91	R/S	Enter $\Delta\lambda$	066	09	9	
026	36	PGM		067	00	0	
027	10	10		068	00	0	
028	12	B		069	65	×	
029	00	0		070	43	RCL	
030	36	PGM		071	27	27	
031	10	10		072	95	=	
032	13	C		073	35	1/X	τ_r
033	91	R/S	Enter O.D.$_i$	074	91	R/S	
034	36	PGM					
035	10	10					
036	91	R/S					
037	61	GTO					
038	00	00					
039	33	33					
040	91	R/S					

Register Contents, Labels, and Data Cards--Algebraic System

Register	Contents	Labels	Contents
R01	Pointer	Label A'	Enter λ_{max}
R02	Counter	Label B'	Enter refractive index
R03	$\Delta\lambda$	Label E	τ_r
R04	1		
R05	h		
R06 $\rightarrow$ Rn*	O.D.$_i$		
R26	$\bar{\nu}_0^2$		
R27	n^2		

*Number of data register available.

Example

Below are the data which represent the absorbance spectrum of aurovertin, a fluorescent inhibitor of the enzyme adenosine triphosphatase of mitochondria. The concentration of aurovertin used to obtain this spectrum was 1.516 μM in 10 mM Tris buffer (pH 7.5). The extinction coefficient for aurovertin at λ_{max} = 367 nm is 28.5 mM^{-1} cm^{-1} (2). Calculate the radiative fluorescence program.

λ (nm)	O.D.
420	4.0×10^{-3}
410	4.4×10^{-3}
400	0.012
390	0.024
380	0.033
370	0.042
360	0.043
350	0.038
340	0.029
330	0.019
320	0.011
310	0.003

Solution

$$\tau_r = 1.980 \times 10^{-9} \text{ sec}$$
$$= 1.980 \text{ nsec}$$

refractive index = 1.33

References

1. C.A. Parker (1968), *Photoluminescence of Solutions*, Elsevier, Amsterdam, pp. 22-28.

2. J.L.M. Muller, J. Rosing, and E.C. Slater (1977), "The Binding of Aurovertin to Isolated F_1(Mitochondrial ATPase)," *Biochimica et Biophysica Acta 462*, 422.437.

3. E.J. Bowen and F. Wokes (1943), *Fluorescence of Solutions*, Longmans Green & Co., London.

6D. FLUORESCENCE POLARIZATION MEASUREMENTS

The steady-state fluorescent measurements of extrinsic fluorescent probes in biological systems can reveal information on the state of the immediate molecular domain. Fluorophores are usually chemically derivatized so that the probe will partition into the structure being studied. For example, the fluorescent probe 12-(9-anthroyl) stearic acid is a fatty acid derivative which partitions into hydrophobic areas of biological membranes. The motion of the molecular probe in the membrane can be determined by the way the fluorescent molecule emits partially polarized light. The degree of polarization (equation 1) of the molecular probe in the medium can be determined by measuring the fluorescence intensities through polarizers which are oriented in the perpendicular $I_\perp$ and parallel $I_\parallel$ to the polarization of the excitation beam (Figure 6D):

$$P = \frac{I_\parallel - I_\perp G}{I_\parallel + I_\perp G} \tag{1}$$

Here G is a correction factor for the unequal transmission of differently polarized light. Probe molecular motion is often expressed as the fluorescence anisotropy r and is taken from the same steady state measurements as the polarization:

$$r = \frac{I_\parallel - I_\perp G}{I_\parallel + 2I_\perp G} \tag{2}$$

If the molecule is spherical and is in an isotropic medium, the rate of probe rotation R can be estimated using the Perrin relationship

$$R = \frac{r_0/r - 1}{6T} \tag{3}$$

In equation 3, r is the measured fluorescence anisotropy, r_0 is the anistropy in the absence of motion, and the ratio of r_0 to r is termed the *degree of depolarization*. The fluorescent lifetime τ is measured separately using a pulsed flashlamp and a gated photomultiplier capable of resolving nanosecond fluorescent decay.

The calculator program includes the input of polarization intensity data, the fluorescent lifetime (a separate measurement), the gating correction factor, and the anisotropy r_0, which is usually the probe anisotropy at -50°C in propylene glycol. Program outputs allow the option of polarization or anisotropy calculations, as well as the rotational motion.

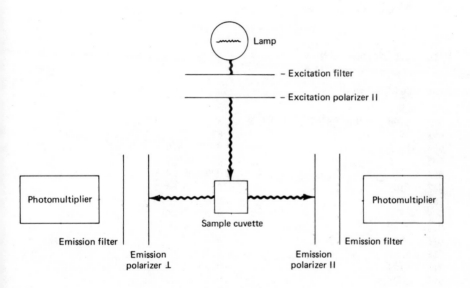

Figure 6D.   Typical polarization measurement geometry.

User Instructions--RPN

Step	Instructions	Input	Keys	Output
1	Enter correction factor	G	f a	
2	Enter measured lifetime (nsec)	τ	f b	
3	Enter r_0	r_0	f c	
4	Enter $I_\perp$	$I_\perp$	ENTER↑	
5	Enter $I_{\parallel}$; compute r	$I_{\parallel}$	A	r
6	Compute P		B	P
7	Compute rate of rotation, R		C	R

Program Listing--RPN

Line	Key	Comments	Line	Key	Comments
001	*LBLa		023	+	
002	ST00	Correction factor G	024	÷	
003	RTN		025	ST05	
004	*LBLb		026	RTN	$r = \dfrac{(I_{\parallel} - I_\perp G)}{(I_{\parallel} + 2I_\perp G)}$
005	6		027	*LBLB	
006	×	→ 6τ	028	RCL4	
007	ST01		029	RCL3	
008	RTN		030	-	
009	*LBLc		031	RCL4	
010	ST02	r_0	032	RCL3	$P = \dfrac{(I_{\parallel} - I_\perp G)}{(I_{\parallel} + I_\perp G)}$
011	RTN		033	+	
012	*LBLA		034	÷	
013	ST04		035	RTN	
014	X⇄Y		036	*LBLC	
015	RCL0	$I_{\parallel}$	037	RCL2	
016	×		38	RCL5	
017	ST03		039	÷	
018	-	$I_\perp G$	040	1	$R = [(r_0/r) - 1]/6\tau$
019	RCL4		041	-	
020	2		042	RCL1	
021	RCL3		043	÷	
022	×		044	RTN	

Register Contents, Labels, and Data Cards--RPN

Register	Contents	Labels	Contents
R_0	G	C	$\rightarrow R$
R_1	6τ	A	$I_\perp \uparrow I_\parallel \rightarrow r$
R_2	r_0	B	$\rightarrow P$
R_3	$I_\perp G$	a	$G\uparrow$
R_4	$I_\parallel$	b	$\tau\uparrow$
R_5	r	c	$r_0\uparrow$

User Instructions--Algebraic System

Step	Instructions	Input	Keys	Output
1	Enter correction factor	G	2nd A'	
2	Enter lifetime	τ	2nd B'	6τ
3	Enter r_0	r_0	2nd C'	
4	Enter $I_\perp$	$I_\perp$	A	$GI_\perp$
5	Enter $I_\parallel$; calculate anistropy	$I_\parallel$	B	r
6	Calculate polarization		C	P
7	Calculate probe rotation		D	R

Program Listing--Algebraic System

Line	Key	Entry	Comments	Line	Key	Entry	Comments
000	76	LBL		041	43	RCL	
001	16	A'	Enter G	042	05	05	
002	42	STO		043	85	+	
003	00	00		044	53	(	
004	91	R/S		045	02	2	
005	76	LBL		046	65	×	
006	17	B'	Enter τ	047	43	RCL	
007	42	STO		048	04	04	Calculate
008	01	01		049	54	)	$\rightarrow r$
009	65	×		050	95	=	
010	06	6	Calculate	051	42	STO	
011	95	=	$\rightarrow 6\tau$	052	06	06	Derive P
012	42	STO		053	91	R/S	
013	01	01		054	76	LBL	
014	91	R/S	Enter r_0	055	13	C	
015	76	LBL		056	53	(	
016	18	C'		057	43	RCL	
017	42	STO	Enter $I_\perp$	058	05	05	
018	02	02		059	75	-	
019	91	R/S		060	43	RCL	
020	76	LBL		061	04	04	
021	11	A		062	54	)	
022	42	STO		063	55	÷	
023	03	03	Calculate	064	53	(	
024	65	×	$\rightarrow I_\perp\ G$	065	43	RCL	
025	43	RCL		066	05	05	
026	00	00		067	85	+	Output
027	95	=		068	43	RCL	$\rightarrow P$
028	42	STO	Enter $I_\parallel$	069	04	04	
029	04	04		070	54	)	
030	91	R/S		071	95	=	Derive R
031	76	LBL		072	91	R/S	
032	12	B		073	76	LBL	
033	42	STO		074	14	D	
034	05	05		075	53	(	
035	75	-		076	53	(	
036	43	RCL		077	43	RCL	
037	04	04		078	02	02	
038	54	)		079	55	÷	
039	55	÷		080	43	RCL	
040	53	(		081	06	06	

Line	Key	Entry	Comments	Line	Key	Entry	Comments
082	54	)		087	43	RCL	
083	75	-		088	01	01	Output R
084	01	1	Output R	089	95	=	
085	54	)		090	91	R/S	
086	55	÷					

Register Contents, Labels, and Data Cards--Algebraic System

Register	Contents	Labels	Contents
R0	G	Label A'	G
R01	6τ	Label B'	τ
R02	r_0	Label C'	r_0
R03	$I_\perp$	Label A	Enter $I_\perp$
R04	$GI_\perp$	Label B	Enter $I_\parallel$; calculate r
R05	$I_\parallel$	Label C	Calculate P
R06	r	Label D	Calculate R

Example

The fluorescent probe 12-(9-anthroyl)stearic acid was added to a suspension of red blood cells. Using the intensity data, determine the anisotropy, polarization, and probe rotation rate as a function of temperature

Temperature (°C)	Relative Intensities		Lifetime, τ (nsec)
	$I_\perp$	$I_\parallel$	
25	7.55	10	12.7
37	6.50	8	12.7
46	5.10	6	10.6

$G = 0.9$
$r_0 = 0.275$

Solution

(°C)	r	P	$R \times 10^7 \text{ sec}^{-1}$
25	0.136	0.191	1.344
37	0.109	0.155	1.994
46	0.093	0.133	3.083

References

1. M. Shinitzky, A. Dianoux, C. Gitler, and G. Weber (1971), *Biochemistry 10*, 2106-2113.

2. J. Vanderkooi, S. Fischkoff, B. Chance, and R. Cooper (1974), *Biochemistry 13*, 1589-1595.

6E.   CIRCULAR DICHROISM AND ROTATIONAL STRENGTH

Some, but not all, biological substances exhibit what is called optical activity. This is manifested in two seemingly different but actually closely related phenomena: circular dichroism (C.D.) and optical rotation.   If circularly polarized light is passed through a solution of an optically active substance, we find that absorptivity depends on the handedness of the light, that is, whether it is left or right circularly polarized.   As shown in Figure 6E, some substances absorb more strongly left circularly polarized light, whereas others may absorb more strongly right polarized light.

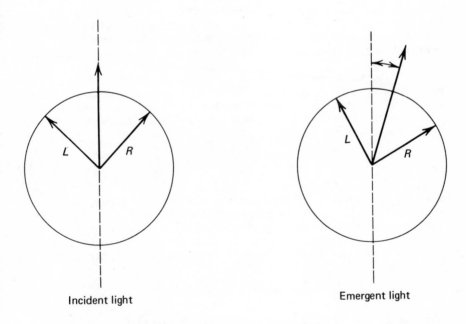

Incident light   Emergent light

Figure 6E.   Differential absorption of left and right circularly polarized light.

The definition of circular dichroism at a given wavelength λ is $\Delta\varepsilon = \varepsilon_L - \varepsilon_R$, the difference in extinction coefficients. Some commercially available instruments measure instead the ellipticity θ; θ is related to $\Delta\varepsilon$ by the equation

$$\theta = 3300\Delta\varepsilon \tag{1}$$

A curve showing the dependence of θ on wavelength is called a *C.D. curve* or *spectrum*. For most work, the terms *molar ellipticity* and *mean residue ellipticity* are used

$$[\theta]_\lambda = \frac{M\theta_\lambda}{10dc} \tag{2}$$

in which θ_λ is the observed ellipticity (degrees), M is the molecular weight or mean residue weight, d is the pathlength (cm), and C is the concentration (gm/ml). The mean residue weight is equal to the molecular weight of the protein divided by the number of amino acid residues. The dimensions of $\Delta\varepsilon$ are liters per centimeter per mole, and $[\theta]$ is in degrees per square centimeter per decimole.

A C.D. band can be characterized by the height, that is, the magnitude of $[\theta]$ at λ_0, which is to some extent a measure of the degree of asymmetry. However, with C.D. it is more common to refer to the rotational strength R of a band. The area under the $\Delta\varepsilon$ versus λ curve is

$$R = \frac{(2.303)(3000)hc}{32\pi^3 N} \int \frac{\Delta\varepsilon}{\lambda} \, d\lambda \tag{3}$$

where h is Planck's constant, c the velocity of light, and N Avogadro's number. It is important to note that circular dichroism is observed only in the wavelength regions where the substance absorbs light. Rotational strength is difficult to evaluate precisely and is normally approximated for a Gaussian band as

$$R \approx 1.23 \times 10^{-42} \, [\theta] \, \frac{\Delta}{\lambda_0} \tag{4}$$

where Δ is the width of the C.D. band at $1/e$ times its height.

The rotational strength describes the intensity of a C.D. band and physically tells something about the motion of electrons when the absorbing center is raised from the ground state to an excited state by the absorption of light; it is not necessary to understand in detail the factors that determine the magnitude of R in order to interpret C.D. data. The main rule is that R is not zero in an optically active substance and that it generally increases with increasing asymmetry.

User Instructions--RPN

Step	Instructions	Input	Keys	Output
1	Enter mean residue weight	M.R.W.	ENTER↑	
2	Enter concentration (gm/ml)	C	ENTER↑	
3	Enter pathlength (cm)	d	f a	
4	Enter scale factor, [e.g., 10^{-4} degree/mm]	Scale	f b	
5	Enter deflection (mm) at wavelength λ	mm_λ	ENTER↑	
6	Enter baseline at wavelength λ (mm)	$base_\lambda$	A	$[\theta]$
	Molar ellipticity $[\theta]$ will be displayed rounded off to the nearest 100			
7	Optional: If Lorentz correction for refractive index is required, press f c		f c	$[\theta]'$
8	If you wish to use a different refractive index, store in R_4; default refractive index is that of water			
	To calculate rotational strength:			
9	Enter: wavelength at maximum $[\theta]$	λ_0	ENTER↑	
	molar ellipticity at λ	θ_0	C	$1/e[\theta]$
10	Enter width of the band at $1/e$ the height of $[\theta]$ (nm)	Δ	D	
11	Calculate rotational strength, R		E	R

Program Listing--RPN

Line	Key	Comments
001	*LBLa	
002	STO2	→ Store d
003	R↓	
004	STO1	→ Store C
005	P↓	
006	STO0	→ Store M.R.W.
007	1	
008	.	
009	3	
010	3	→ Store n,
011	STO4	refractive index
012	CLX	of water
013	RTN	
014	*LBLA	
015	-	(mm - base) x scale
016	RCL3	$= \theta_{obs}$
017	RCL0	
018	X	Compute
019	X	$\theta = \dfrac{\theta_{obs} \text{M.R.W.}}{10dC}$
020	1	
021	0	
022	RCL2	
023	X	
024	RCL1	
025	X	
026	÷	
027	DSP0	
028	RND	Round off and
029	DSP2	print to nearest
030	STOA	100
031	PRTX	
032	RTN	
033	*LBLb	
034	STO3	→ Store scale
035	RTN	
036	*LBLC	
037	STO6	→ Store θ
038	R↓	
039	STO5	→ Store λ_0
040	1	
041	e^x	
042	1/X	
043	RCL6	Calculate $1/e$ x θ
044	X	
045	DSP0	

Line	Key	Comments
046	RND	Round off and dis-
047	DSP2	play to nearest 100
048	RTN	
049	*LBLD	→ Store Δ
050	STO7	
051	RTN	
052	*LBLE	Compute rotational
053	RCL7	strength
054	RCL5	
055	÷	
056	RCL6	
057	X	
058	1	
059	.	
060	2	
061	3	
062	EEX	$R \approx 1.23 \times 10^{-42} [\theta] \dfrac{\Delta}{\lambda_0}$
063	4	
064	2	
065	CHS	
066	X	
067	PRTX	
068	RTN	
069	*LBLc	
070	RCL4	
071	X^2	
072	2	
073	+	
074	3	
075	X≠Y	
076	÷	Lorentz correction
077	RCLA	for refractive
078	X	index
079	DSP0	
080	RND	
081	DSP2	
082	PRTX	
083	RTN	

Register Contents, Labels, and Data Cards--RPN

Register	Contents	Labels	Contents
R_0	M.R.W.	A	$mm_\lambda \uparrow base_\lambda \rightarrow [\theta]$
R_1	C	C	$\lambda_0 \uparrow \theta_0$
R_2	d	D	$\Delta\uparrow$
R_3	Scale	E	$\rightarrow R$
R_4	n	a	M.R.W. $\uparrow C \uparrow d$
R_5	λ_0	b	Scale$\uparrow$
R_6	θ_0	c	Lorentz correction
R_7	Δ		
R_A	$[\theta]$		

User Instructions--Algebraic System

Step	Instructions	Input	Keys	Output
1	Enter pathlength	d	2nd A'	
2	Enter concentration	C	2nd B'	
3	Enter mean residue weight	M.R.W.	2nd C'	
4	Enter scale	Scale	2nd D'	
5	Optional: enter refractive index other than water	n	R/S	
6	Optional: Lorentz correction for refractive index		R/S	$[\theta]'$
7	Enter measurement	measurement	A	

Step	Instructions	Input	Keys	Output
8	Enter baseline; calculate θ	baseline	R/S	[θ]
9	Enter θ_0; calculate $1/e(\theta_0)$	θ_0	B	θ_0
10	Enter λ_0	λ_0	R/S	
11	Enter Δ; calculate rotational strength	Δ	C	R

Program Listing--Algebraic System

Line	Key	Entry	Comments	Line	Key	Entry	Comments
000	76	LBL		029	20	20	
001	16	A'	Enter d	030	91	R/S	
002	42	STO		031	76	LBL	Enter measurement
003	00	00		032	11	A	
004	91	R/S		033	42	STO	
005	76	LBL		034	04	04	
006	17	B'	Enter C	035	91	R/S	
007	42	STO		036	42	STO	Enter baseline
008	01	01		037	05	05	
009	91	R/S		038	53	(	
010	76	LBL	Enter M.R.W.	039	43	RCL	
011	18	C'		040	04	04	
012	42	STO		041	75	-	
013	02	02		042	43	RCL	$\theta_\lambda = \dfrac{m\theta_{obs}}{10dC}$
014	91	R/S		043	05	05	
015	76	LBL	Enter scale	044	54	)	
016	19	D'		045	65	$\times$	
017	42	STO		046	43	RCL	
018	03	03		047	03	03	
019	01	1		048	95	=	
020	93	.		049	42	STO	
021	03	3		050	06	06	
022	03	3		051	53	(	
023	33	X²		052	43	RCL	
024	42	STO	Store refractive index	053	02	02	
025	20	20		054	65	$\times$	
026	91	R/S		055	43	RCL	
027	33	X²		056	06	06	
028	42	STO		057	54	)	

Line	Key	Entry	Comments	Line	Key	Entry	Comments
058	55	÷		100	04	4	
059	53	(		101	02	2	
060	01	1		102	94	+/-	
061	00	0		103	65	×	
062	65	×		104	43	RCL	
063	43	RCL		105	08	08	
064	00	00		106	65	×	
065	65	×		107	53	(	
066	43	RCL		108	43	RCL	
067	01	01		109	10	10	
068	54	)		110	55	÷	
069	95	=	Round off	111	43	RCL	
070	42	STO	$[\theta_\lambda]$	112	09	09	
071	07	07		113	54	)	
072	91	R/S		114	95	=	Compute $[\theta]'$
073	61	GTO		115	91	R/S	
074	39	COS		116	76	LBL	
075	76	LBL		117	39	COS	
076	12	B		118	03	3	
077	22	INV		119	55	÷	
078	58	FIX		120	53	(	
079	42	STO		121	43	RCL	
080	08	08		122	20	20	
081	65	×		123	85	+	
082	93	.		124	02	2	
083	03	3		125	54	)	
084	06	6	$1/e[\theta_0]$	126	65	×	
085	08	8		127	43	RCL	$[\theta]'$
086	95	=		128	07	07	
087	91	R/S		129	95	=	
088	42	STO		130	91	R/S	
089	09	09	Enter λ_0				
090	91	R/S	Compute R				
091	76	LBL					
092	13	C					
093	42	STO					
094	10	10					
095	01	1					
096	93	.					
097	02	2					
098	03	3					
099	52	EE					

Register Contents, Labels, and Data Cards--Algebraic System

Register	Contents	Labels	Contents
R01	d	Label A'	Enter d
R01	c	Label B'	Enter c
R02	M.R.W.	Label C'	Enter M.R.W.
R03	Scale	Label D'	Enter scale
R04	Measurement	Label A	Enter measurement
R05	Baseline	Label B	Enter $\theta \rightarrow 1/e[\theta_0]$
R06	θ_{obs}	Label C	Enter $\Delta \rightarrow R$
R07	$[\theta_\lambda]$	Label D	Derive $[\theta]'$
R08	θ_0		
R09	λ_0		
R10	Δ		
R20	n		

Example

Below are the data from a C.D. spectrum of the enzyme adenosine triphosphatase from mitochondria. The protein concentration was 102 μg/ml in a 0.1 mm pathlength cell. The buffer used was 10 mM Tris (pH 7.5). The scale was 1 millidegree/cm or 10^{-4} degrees/mm.
The mean residue weight of mitochondrial ATPase is 119.

λ (nm)	Deflection (mm)	Baseline (mm)	[θ]
218	97.5	209	−13,008
219	95.5	207	−13,008
220	95	206	−12,950
221	93.5	200	−12,425
222	94	208	−13,300
223	93.5	204	−12,892
224	92	198	−12,367
225	90.5	194.5	−12,133
226	87.5	187.5	−11,667
227	83	178	−11,083
228	76.5	164	−10,208

References

1. David Freifelder (1976), *Physical Biochemistry--Applications to Biochemistry and Molecular Biology*, W.H. Freeman and Company, San Francisco, pp. 449-454.

2. K.E. Van Holde (1971), *Physical Biochemistry*, Foundation in Modern Biochemistry Series, Prentice-Hall, Englewood Cliffs, N.J., pp. 205-210.

6F. CALCULATION OF SECONDARY STRUCTURES OF PROTEINS FROM CIRCULAR DICHROISM

The rotatory contributions of a protein can be represented by

$$[\theta] = [\theta]_H f_H + [\theta]_\beta f_\beta + [\theta]_R f_R \qquad (1)$$

The f's are the fractions of the helix (H), the β form, and the unordered or random coil form (R); $f_H + f_\beta + f_R = 1$ and all f's ≥ 0. The $[\theta]_H$, $[\theta]_\beta$, and $[\theta]_R$ are the reference values that would be obtained if the protein molecule were made up of segments of pure helix, β form, and unordered (random coil) form.

Solving for the θ parameters in equation 1 requires a minimum of three simultaneous equations with the use of the experimental values for three proteins whose f values can be deduced from X-ray diffraction studies. Chen, Yang, and Martinez (3) chose five model proteins (see Table 2) from which the computed the $[\theta]$ values for pure helix, β, and random coil structures.

The resulting values of $[\theta]$ for helix, β, and R forms thus determined can then be used conversely to determine the f_H, f_β, and f_R of any protein, using $[\theta]$'s at three wavelengths by solving three equations in three unknowns. Barela and Darnall (4), on the basis of a more thorough statistical analysis of the calculations of protein secondary structure from C.D. spectra, concluded that the 220-240 nm segment of the spectrum was the region of greatest error. The greatest uncertainty associated with this method was the choice of the experimental wavelengths to be used in the analysis, and the fact that no choice of wavelengths could be made which would be applicable to all proteins. These authors urged that a great deal of caution be exercised before an interpretation of the C.D. spectrum of an unknown protein is made, and that the common practice of calculating the amount of helix in a protein form ellipticity measurements at 222 nm be discontinued.

With this in mind, we recommend that structure fractions be calculated at several sets of three wavelengths throughout the region of 210-230 nm. Although this procedure is still only an approximation, the structure fractions so obtained should provide some insight into the secondary structure of protein in solution

The sample problem provided here is one in which the actual experimental data of one of the reference proteins in Table 1 (ribonuclease) are used to compute the structure fractions, using either the 3 × 3 matrix operations program from the Hewlett-Packard Standard Pac (SD-10A) or the simultaneous equations program in the Master Library Crom with the TI-58/59.

TABLE 1. FRACTIONS OF HELIX, β, AND RANDOM COIL FORMS OF FIVE
PROTEINS AS DETERMINED BY X-RAY DIFFRACTION STUDIES

Protein	f_H	f_β	f_R
Myoglobin	0.77	0	0.23
Lysozyme	0.29	0.16	0.55
Lactate dehydrogenase	0.29	0.20	0.51
Papain	0.21	0.05	0.74
Ribonuclease	0.19	0.38	0.43

TABLE 2. CIRCULAR DICHROISM DATA OF HELIX, β, AND RANDOM COIL
FORMS, BASED ON FIVE REFERENCE PROTEINS, (207-237 nm)[*]

λ (nm)	$[\theta]$		
	H	β	R
207	-22,288	-4,317	-5,770
210	-26,437	-8,189	-2,203
213	-24,822	-8,679	-848
216	-26,554	-9,213	+1,229
219	-28,881	-6,890	+1,723
222	-30,044	-3,361	+1,581
225	-28,664	+1,542	+264
228	-24,030	+4,394	-480
231	-17,322	+4,575	-774
234	-11,337	+3,538	+163
237	-6,248	+2,405	-88

*These data were generously provided by Dr. J.T. Yang, Cardio-
vascular Research Institute and Department of Biochemistry and
Biophysics, University of California, San Francisco.

321

Example

By using data for the protein ribonuclease obtained from the paper by Chen et al. (3), a set of three equations in three unknowns can be set up; these take the following form:

$$[\theta]_{216} = [\theta_{H216}]f_H + [\theta_{\beta216}]f_\beta + [\theta_{R216}]f_R$$

$$[\theta]_{219} = [\theta_{H219}]f_H + [\theta_{\beta219}]f_\beta + [\theta_{R219}]f_R$$

$$[\theta]_{222} = [\theta_{H222}]f_H + [\theta_{\beta222}]f_\beta + [\theta_{R222}]f_R$$

at wavelengths of 216, 219, and 222 nm.

Using the actual values of θ_H, θ_β, and θ_R from Table 2 gives

$$-10,100 = (-26,554)f_H + (-9,213)f_\beta + (1,229)f_R$$

$$-9,420 = (-28,881)f_H + (-6,890)f_\beta + (1,723)f_R$$

$$-8,090 = (-30,044)f_H + (-3,361)f_\beta + (1,723)f_R$$

Solving three equations in three unknowns, using the 3 × 3 matrix operations program from the Hewlett-Packard Standard Pac (SD-10A) or from the TI-58/59, Master Library Program 02 yields the following results:

Fraction	Calculated	X-ray Values
f_H	0.23	0.19
f_β	0.47	0.38
f_R	0.29	0.43

For another set of three wavelengths (219, 222, and 225 nm):

$$-9,420 = (-28,881)f_H + (-6,890)f_\beta + (1,723)f_R$$

$$-8,090 = (-30,044)f_H + (-3,361)f_\beta + (1,581)f_R$$

$$-6,130 = (-28,664)F_H + (1,542)f_\beta + (264)f_R$$

The calculated results are as follows:

Fraction	Calculated	X-ray Values
f_H	0.24	0.19
f_β	0.48	0.38
f_R	0.55	0.43

Clearly, the reliability of this calculation is quite poor since in some cases the sum of the fractions exceeds 1.00. The helical structure fraction value is the only one that comes reasonably close to the X-ray structure value, and it remains relatively constant independently of the other structure fractions (β and R) through the wavelength region of 216-228 nm.

References

1. K.E. Van Holde (1971), *Physical Biochemistry,* Foundations in Modern Biochemistry Series, Prentice-Hall, Englewood Cliffs, N.J., Chapter 10.

2. David Freifelder (1976), *Physical Biochemistry--Applications to Biochemistry and Molecular Biology,* W.H. Freeman and Company, San Francisco, Chapter 16.

3. Y.-H. Chen, J.T. Yang, and H.M. Martinez (1972), "Determination of the Secondary Structures of Proteins by Circular Dichroism and Optical Rotatory Dispersion," *Biochemistry 11,* 4120-4131.

4. T.D. Barela and D.W. Darnall (1974), "Practical Aspects of Calculating Protein Secondary Structure from Circular Dichroism Spectra," *Biochemistry 13,* 1394-1400.

VII

ISOTOPES IN BIOCHEMISTRY

7A. RADIOISOTOPE DECAY

The decay of radioactive isotopes is a simple exponential (first-order) process:

$$- \frac{dN}{dt} = \lambda N \tag{1}$$

where $-dN/dt$ = number of atoms decaying per small increment of time (i.e., the count rate)

N = total number of radioactive atoms present at any given time

λ = a decay constant, different for each isotope

The negative sign indicates that the number of radioactive atoms decreases with time.

Although λ is a proportionality constant, the physical significance is seen by rearranging equation 1:

$$\lambda = - \frac{dN/N}{dt} \tag{2}$$

Therefore λ is the fraction of the radioactive atoms that decays per small increment of time.

Equation 1 can be integrated between the limits of N_o (the original number of radioactive atoms) and N (the number of radioactive atoms at any other time) and between the limits of zero time and any other time:

$$\ln \frac{N_o}{N} = \lambda t \tag{3}$$

or

$$N = N_o e^{-\lambda t} \tag{4}$$

The terms N_O and N can be expressed in any consistent manner. For example, N_O = 100%, N = percent remaining after time interval t; N_O = 1.00, N = fraction remaining (as a decimal) after time interval t; N_O = original count per minute in sample, N = count per minute remaining after time interval t; N_O = initial millicuries of sample, N = millicuries of sample after an elapsed time t.

The half-life $t_{1/2}$ of a radioactive isotope is the time required for half of the original atoms to decay. The relationship between $t_{1/2}$ and λ is as follows:

$$\lambda = \frac{0.693}{t_{1/2}} \qquad \text{or} \qquad t_{1/2} = \frac{0.693}{\lambda} \tag{5}$$

Half-lives for some of the isotopes commonly used in biological studies are given in Table 1.

Table 1

Isotope	$t_{1/2}$
^{125}I	60 days
^{131}I	8.1 days
^{3}H	12.3 days
^{14}C	5700 years
^{32}P	14.3 days
^{35}S	87.1 days
^{45}Ca	163 days

Radioactivity is expressed in units of curies. One curie
(Ci) is defined as the number of disintegrations per second per
gram of radium and equals 3.7 x 10^{10} disintegrations per second.
For most biological applications, quantities much less than 1 Ci
are normally used and the milli- (mCi) or microcurie (μCi) is em-
ployed.   In practice, a minute is the standard time unit; therefore 1 μCi = 2.22 x 10^6 disintegrations per minute (dpm).   Because the efficiency of most radiation detection devices is less
than 100%, a given number of curies almost always yields a lower
than theoretical count rate.   Hence there is a distinction between dpm and cpm (counts per minute).   For example, a sample
containing 1 μCi of radioactive material has a decay rate of
2.22 x 10^6 dpm.   If only 30% of the disintegrations are detected,
the observed count rate is 6.66 x 10^5 cpm.

The program presented in this section allows the user to
determine the remaining radioactivity of a sample by entering the
initial date as a decimal number (in the format MM.DDYYYY for the
HP-67 and MMDD.YYYY for the TI-58/59) the second date in the same
format, followed by the radioactivity on the initial date in
either millicuries or counts per minute.   The remaining activity
will be calculated and displayed.   The user must first enter the
half-life (in days) of the radioisotope in question and then compute the number of days between dates 1 and 2.

User Instructions--RPN

Step	Instructions	Input	Keys	Output
1	Enter half-life, $t_{\frac{1}{2}}$, of isotope	$t_{\frac{1}{2}}$	f a	$t_{\frac{1}{2}}$
2	Enter date 1 as MM.DDYYYY	Date 1	A	Julian Day 1
3	Enter date 2 as MM.DDYYYY	Date 2	B	Julian Day 2
4	Calculate number of days between date 1 and date 2		f b	Δ days
	Or, if Δ days is already known, store Δ days directly in R_B	Δ days	Sto B	Δ days
5	Enter mCi on Date 1 and compute mCi remaining on date 2	mCi_1	C	mCi_2
6	Enter cpm on date 1 and compute cpm remaining on date 2	cmp_1	D	cpm_2

Program Listing--RPN

Line	Key	Comments	Line	Key	Comments
001	*LBLA		015	3	
002	3	Control 3 in	016	0	
003	GTO0	display	017	.	
004	*LBLB		018	6	
005	4	Control 4 in	019	0	
006	*LBL0	display	020	0	
007	STOI		021	1	
008	3		022	STO6	
009	6		023	R↑	
010	5	Store constants	024	ENT↑	Break input date
011	.		025	INT	into individual com-
012	2		026	STO7	ponents of MM,DD,
013	5		027	-	YYYY
014	STO5		028	EEX	

Line	Key	Comments	Line	Key	Comments
029	2		075	x	
030	x		076	-	
031	ENT↑		077	RTN	
032	INT		078	*LBLb	Calculate Δ days
033	STO8		079	DSP0	
034	-		080	RCL4	
035	EEX		081	RCL3	
036	4		082	-	
037	x		083	STOB	
038	STO9		084	RTN	
039	RCL7		085	*LBLa	Store half-life of
040	1		086	DSP2	isotope (days)
041	+		087	STOA	
042	ENT↑		088	RTN	
043	1/X		089	*LBLC	
044	.		090	DSP2	Compute remaining
045	7		091	ENT↑	activity:
046	+		092	RCLB	$mCi_1 \rightarrow mCi_2$
047	CHS		093	RCLA	
048	GSB2	Compute day number	094	÷	
049	RCL6		095	.	
050	x		096	5	
051	INT		097	X⇄Y	
052	RCL9		098	Yˣ	
053	RCL5		099	x	
054	x		100	RTN	
055	INT		101	*LBLD	Compute remaining
056	+		102	GSBC	activity:
057	RCL8		103	DSP0	$cpm_1 \rightarrow cpm_2$
058	+		104	RTN	
059	STO i	Compute Julian Day			
060	1	number for output			
061	7				
062	2				
063	0				
064	9				
065	8				
066	2				
067	+				
068	DSP0				
069	RTN				
070	*LBL2				
071	INT				
072	ST+9				
073	1				
074	2				

Register Contents, Labels, Data Cards--RPN

Register	Contents	Labels	Contents
R_3	Julian Day 1	A	Date 1 $\uparrow$
R_4	Julian Day 2	B	Date 2 $\uparrow$
R_5	365.25	C	$mCi_1 \rightarrow mCi_2$
R_6	30.6001	D	$cpm_1 \rightarrow cpm_2$
R_7	MM	a	$t_{\frac{1}{2}}$ (days)
R_8	DD	b	$\rightarrow$ days
R_9	YYYY		
R_A	$t_{\frac{1}{2}}$ (days)		
R_B	Δ days		
R_I	Control		

User Instructions--Algebraic System

Step	Instructions	Input	Keys	Output
1	Enter half-life, $t_{\frac{1}{2}}$, of isotope	$t_{\frac{1}{2}}$	2nd A'	$t_{\frac{1}{2}}$
2	Enter date 1 as MMDD.YYYY	Date 1	A	0
3	Enter date 2 as MMDD.YYYY; calculate number of days between date 1 and date 2	Date 2	B	Δ days
	Or, if Δ days is already known, enter directly	Δ days	D	
4	Enter mCi on date 1 and compute mCi remaining on date 2	mCi_1	E	mCi_2
5	Enter cpm on date 1 and compute cpm remaining on date 2	cpm_1	2nd E'	cpm_2

330 7. Isotopes in Biochemistry

Program Listing--Algebraic System

Line	Key	Entry	Comments	Line	Key	Entry	Comments
000	76	LBL		041	06	06	
001	16	A'	Enter half-	042	65	×	
002	42	STO	life	043	43	RCL	
003	06	06		044	09	09	
004	91	R/S		045	94	+/-	
005	76	LBL	Enter date 1	046	95	=	
006	11	A		047	22	INV	
007	42	STO	Call module	048	23	LNX	
008	07	07	date sub-	049	65	×	
009	36	PGM	routine	050	43	RCL	
010	20	20		051	10	10	
011	11	A		052	95	=	
012	92	RTN	Enter date 2	053	91	R/S	
013	76	LBL		054	76	LBL	Enter cpm at
014	12	B		055	10	E'	date 1
015	42	STO		056	42	STO	
016	08	08		057	11	11	
017	36	PGM	Call module	058	93	.	
018	20	20	date sub-	059	06	6	
019	12	B	routine	060	09	9	
020	36	PGM		061	03	3	Calculate cpm
021	20	20		062	55	÷	remaining at
022	13	C	Calculate	063	43	RCL	date 2
023	42	STO	days between	064	06	06	
024	09	09	dates	065	65	×	
025	92	RTN		066	43	RCL	
026	76	LBL	If number of	067	09	09	
027	14	D	days is	068	94	+/-	
028	42	STO	known, enter	069	95	=	
029	09	09		070	22	INV	
030	91	R/S		071	23	LNX	
031	76	LBL		072	65	×	
032	15	E	Enter mCi at	073	43	RCL	
033	42	STO	date 1	074	11	11	
034	10	10		075	95	=	
035	93	.		076	91	R/S	
036	06	6	Calculate				
037	09	9	mCi remaining				
038	03	3	at date 2				
039	55	÷					
040	43	RCL					

Register Contents, Labels, and Data Cards--Algebraic System

Register	Contents	Labels	Contents
R1 → R5	Days between dates	Label A'	Enter $t_{\frac{1}{2}}$
R6	$t_{\frac{1}{2}}$	Label A	Enter date 1
R7	Date 1	Label B	Enter date 2
R8	Date 2	Label C	Calculate Δ days
R9	Δ days	Label D	Enter Δ days
R10	mC_i on date 1	Label E	Enter mC_{i_1}
R11	cpm on date 1	Label E'	Enter cpm_1

Example

A. How many millicuries of ^{32}P remain on February 13, 1978, if on February 6, 1978, there were 0.20 mCi?

Solution

7 days elapsed; 0.142 mCi remains.

B. How may counts per minute of ^{35}S remain on February 13, 1978, if on August 15, 1977, there were 3.2×10^6 cpm?

Solution

182 days elapsed, 7.52×10^5 cpm remains.

References

1. I.H. Segel (1976), *Biochemical Calculations*, 2nd Ed., John Wiley & Sons, Inc., New York, pp. 354-358.

2. David Freifelder (1976), *Physical Biochemistry--Applications to Biochemistry and Molecular Biology*, W.H. Freeman and Company, San Francisco, pp. 91-93.

7B. DOUBLE ISOTOPE OVERLAP CORRECTIONS

The β particles from a given radioactive isotope are emitted with a continuous energy distribution extending up to some maximum value (e.g., 0.0176 MeV for ^{3}H, 0.155 MeV for ^{14}C, 1.701 MeV for ^{32}P). When two isotopes have different emission energy spectra, the amount of each present in a mixture can be determined by selectively measuring the radioactivity at different energy levels. This can be accomplished easily with a two-channel scintillation counter. Sometimes it is impossible to attain 100% discrimination, and consequently the activity measured in one channel (or both) will result from both isotopes (Figure 7B). Nevertheless, the amount of each isotope present can still be calculated.

The greatest value of scintillation counting is the ability to determine the ratio of two isotopes present in a mixture. This is possible because the voltage pulse produced as a result of decay is proportional to the energy of the emitted particles. The resolution of the two isotopes is accomplished with a pulse height analyzer equipped with discriminators. The instrument is used in circuitry that is designed to count pulses in different voltage intervals. The discriminators are the controls that determine the voltage levels defining these voltage intervals. A plot of decay rate versus energy or pulse height allows one to determine the amount of spillover or overlap for an isotope counted in one channel compared to that for a different isotope counted in another channel. Figure 7B shows such a plot for ^{3}H and ^{14}C. By using this ability to determine the decay rate in a given pulse height interval, the ratio of two isotopes can be determined.

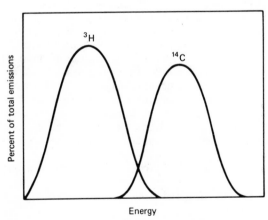

Figure 7B. Overlap of energy distribution spectra of ^{3}H and ^{14}C.

Consider a radioactive sample containing both ^{32}P and ^{35}S
radioisotopes counted in the voltage ranges defined by channels
A and B. (The reader should consult the operator's manual of the
specific scintillation counter involved for instructions concerning how to select the optimum voltage ranges for channels A and
B.) By counting ^{32}P and ^{35}S standard samples, using the same
instrument settings, the amounts of ^{32}P and ^{35}S in each channel
can be calculated as shown in Table 1.

TABLE 1

	Observed cpm (corrected for background)	
Sample	Channel A	Channel B
^{32}P standard	39,000	14,000
^{35}S standard	18,000	56,000
Experimental	74,600	38,500

The background counting rate must be determined independently for
each channel with samples containing no radioisotope. The distribution of counts per minute of the ^{32}P and ^{35}S standard samples
in channels A and B tells how the ^{32}P and ^{35}S are distributed in
an experimental sample. Line 1 of Table 1 shows that the ^{32}P
counts per minute measured in channel B are a result of spillover from channel A and amount to 35.9% of the counts in channel
A. Likewise, line 2 shows that the ^{35}S counts per minute measured in channel A are a result of spillover from channel B and
amount to 32.4% of the counts in channel B.

The calculator program corrects for spillover between channels when two radioisotopes are counted in a liquid scintillation
spectrometer. Background subtraction for each isotope is also
provided. The program may also be used with a single isotope.

Isotopes X and Y are counted in machine channels A and B,
respectively.

Let a = fractional spillover of isotope Y from channel B to
A (B → A)

    b = fractional spillover of isotope X from channel A to
B (A → B)

C_X = corrected counts per minute of isotope X in channel A

$$= \frac{C_A - aC_B}{1 - ab}$$

where C_A and C_B are the observed counts per minute in each channel. Similarly,

$$C_Y = \frac{C_B - bC_A}{1 - ab}$$

Total cpm of isotope $X = T_X = C_X(1 + b)$

Total cpm of isotope $Y = T_Y = C_Y(1 + a)$

This program was submitted by Lawrence I. Grossman to the
Hewlett-Packard User's Library and was published in the Hewlett-
Packard booklet entitled *HP-67/97 User's Library Solutions in
Physics*.

User Instructions--RPN

Step	Instructions	Input	Keys	Output
1	Set flag 0 for two isotopes		f a	0.00
2	Enter number of first sample	n_i	f c	n_i
3	For single isotope: Enter fractional spillover from channel $B \to A$ and background subtraction for channel A	$B \to A$ Bk-A	ENTER↑ D	
4	If two isotopes: Enter fractional spillover from channel $A \to B$ and background subtraction for channel B	$A \to B$ Bk-B	ENTER↑ E	
5	Enter cpm for channel A; for one isotope, go to step 7	cpm-A	A	$C_A - Bk_A$
6	Enter cpm for channel B	cpm-B	B	$C_B - Bk_B$
7	Output corrected cpm for channels A and B (if used)		C	T_X T_Y Next n_i
8	For next sample or pair of samples, repeat steps 5-7			
9	After last entry, obtain total of all samples of isotopes X and Y		f b	ΣT_X ΣT_Y 0.00
	(After outputs are complete, flag for two isotopes is cleared and the accumulation registers for total of samples are cleared)			

Program Listing--RPN

Line	Key	Comments
001	*LBLD	
002	STO0	→ Bk-A
003	R↓	
004	STO1	→ a
005	RTN	
006	*LBLc	
007	STOI	→ First sample
008	DSP0	number
009	PRTX	
010	RTN	
011	*LBLE	
012	STO2	→ Bk- B
013	R↓	
014	STO3	→ b
015	RTN	
016	*LBLA	
017	RCL0	Enter cpm-A;
018	-	subtract Bk-A and
019	STO4	store
020	RTN	
021	*LBLB	
022	RCL2	Enter cpm-B;
023	-	subtract Bk-B and
024	STO5	store
025	RTN	
026	*LBLC	Calculate C_X, C_Y,
027	RCL4	T_X, and T_Y
028	RCL1	
029	RCL5	
030	x	
031	-	
032	1	
033	RCL1	
034	RCL3	
035	x	
036	-	
037	÷	
038	1	
039	RCL3	
040	+	
041	x	
042	ST+6	→ ΣT_X
043	STO7	→ C_X
044	DSP0	

Line	Key	Comments
045	RCL5	
046	RCL3	
047	RCL4	
048	x	
049	-	
050	1	
051	RCL1	
052	RCL3	
053	x	
054	-	
055	÷	
056	1	
057	RCL1	
058	+	
059	x	→ ΣT_Y
060	ST+8	→ C_Y
061	STO9	
062	RCL7	Output T_X, T_Y (if
063	PRTX	flag 0 is set)
064	RCL9	
065	F0?	
066	PRTX	
067	SPC	
068	ISZI	
069	RCLI	
070	PRTX	
071	RTN	
072	*LBLa	
073	SF0	Set flag 0 for two
074	RTN	isotopes
075	*LBLb	
076	SPC	Output T_X for all
077	RCL6	samples, T_Y
078	PRTX	
079	RCL8	
080	F0?	
081	PRTX	
082	CF0	Clear flag 0; clear
083	0	accumulation regis-
084	STO6	ters
085	STO8	
086	STOI	
087	RTN	

Register Contents, Labels, and Data Cards--RPN

Register	Contents	Labels	Contents
R_0	Bk-A	A	cpm-A ↑
R_1	a	B	cpm-B ↑
R_2	Bk-B	C	→ T_X, T_Y
R_3	b	D	$B → A$ ↑ Bk-A
R_4	c_A	E	$A → B$ ↑ Bk-B
R_5	c_B	a	2 isotopes?
R_6	T_X	b	→ ΣT_X, ΣT_Y
R_7	c_X	c	First fraction number
R_8	T_Y		
R_9	c_Y		
R_I	Sample number		

USER INSTRUCTIONS--ALGEBRAIC SYSTEM

Step	Instructions	Input	Keys	Output
1	Enter percent overlap $B \to A$	$B \to A$	2nd A'	
2	Enter percent overlap $A \to B$	$A \to B$	2nd B'	$a \times b$
3	Optional: Enter background A	Bkg A	2nd C'	
4	Optional: Enter backbround B	Bkg B	2nd D'	
5	Enter observed cpm for channel A	c_A	A	
6	Enter observed cpm for channel B	c_B	B	
7	Calculate total cpm for channel A		C	$\sum c_A$
8	Calculate total cpm for channel B			$\sum c_B$

PROGRAM LISTING--ALGEBRAIC SYSTEM

Line	Key	Entry	Comments	Line	Key	Entry	Comments
000	76	LBL		013	42	STO	
001	16	A'	Spillover	014	05	05	
002	42	STO	$B \to A$	015	91	R/S	
003	00	00		016	76	LBL	Bkg A
004	91	R/S		017	18	C'	
005	76	LBL	Spillover	018	42	STO	
006	17	B'	$A \to B$	019	02	02	Flag for Bkg
007	42	STO		020	86	STF	A
008	01	01		021	00	00	Subtraction
009	65	×		022	91	R/S	
010	43	RCL		023	76	LBL	
011	00	00	$a \times b$	024	19	D'	
012	95	=		025	42	STO	

Line	Key	Entry	Comments	Line	Key	Entry	Comments
026	03	03		068	53	(	
027	86	STF	Flag for Bkg	069	43	RCL	
028	01	01	B	070	04	04	
029	91	R/S	Subtraction	071	75	-	
030	76	LBL		072	53	(	
031	11	A	Enter ob-	073	43	RCL	
032	42	STD	served cpm A	074	00	00	
033	04	04		075	65	×	
034	87	IFF		076	43	RCL	
035	00	00	Flag for op-	077	06	06	
036	38	SIN	tional bkg	078	54	)	
037	76	LBL	subtraction	079	54	)	
038	38	SIN		080	55	÷	
039	43	RCL		081	53	(	
040	04	04		082	01	1	
041	75	-		083	75	-	
042	43	RCL		084	43	RCL	
043	02	02		085	05	05	
044	95	=		086	54	)	
045	42	STD		087	95	=	
046	04	04	Enter ob-	088	42	STD	
047	91	R/S	served cpm B	089	07	07	
048	76	LBL		090	65	×	
049	12	B	Flag for op-	091	53	(	C_A total cpm
050	42	STD	tional Bkg	092	01	1	
051	06	06	subtraction	093	85	+	Calculate cor-
052	87	IFF		094	43	RCL	rected cpm
053	01	01		095	01	01	and total cpm
054	39	COS		096	54	)	B
055	76	LBL		097	95	=	
056	39	COS		098	91	R/S	
057	43	RCL		099	76	LBL	
058	06	06		100	14	D	
059	75	-		101	53	(	
060	43	RCL		102	43	RCL	
061	03	03		103	06	06	
062	95	=		104	75	-	
063	42	STD	Calculate	105	53	(	
064	06	06	corrected	106	43	RCL	
065	91	R/S	cpm and total	107	01	01	
066	76	LBL	cpm A	108	65	×	
067	13	C		109	43	RCL	

Line	Key	Entry	Comments		Line	Key	Entry	Comments
110	04	04			121	42	STO	
111	54	)			122	08	08	
112	54	)			123	65	×	
113	55	÷			124	53	(	
114	53	(			125	01	1	
115	01	1			126	85	+	
116	75	-			127	43	RCL	
117	43	RCL			128	00	00	
118	05	05			129	54	)	
119	54	)			130	95	=	Display C_B
120	95	=			131	91	R/S	total cpm

Register Contents, Labels, and Data Cards--Algebraic System

Register	Contents	Labels	Contents
R0	b	Label A'	Enter $B \to A$
R01	a	Label B'	Enter $A \to B$
R02	Bkg A	Label C'	Bkg A
R03	Bkg B	Label D'	Bkg B
R04	C_A	Label A	Enter C_A
R05	$a \times b$	Label B	Enter C_B
R06	C_B	Label C	Calculate $\sum C_A$
R07	$C_{A\ corr}$	Label D	Calculate $\sum C_B$
R08	$C_{B\ corr}$		

Example

Consider a hypothetical experiment using two isotopes where the spillover from channel A to B $(A \to B)$ = 10% and that from channel B to A $(B \to A)$ = 20%.

The background counts per minute in channels A and B are 10 and 50 cpm, respectively.

For the following values of cpm_A and cpm_B, calculate corrected values and totals.

Sample	Channel A	Channel B
1	1000	500
2	2000	1000
3	1400	2200

Solution

Sample 1

$$T_X = 1010$$
$$T_Y = 430$$

Sample 2

$$T_X = 2020$$
$$T_Y = 920$$

Sample 3

$$T_X = 1078$$
$$T_Y = 2462$$

$$\Sigma T_X = 4108$$
$$\Sigma T_Y = 3812$$

References

1. David Freifelder (1976), *Physical Biochemistry--Applications to Biochemistry and Molecular Biology*, W.H. Freeman and Company, San Francisco, pp. 91-103.

2. I.H. Segel (1976), *Biochemical Calculations*, 2nd Ed., John Wiley & Sons, Inc., New York, pp. 373-376.

3. Lawrence I. Grossman (1977), "Isotope Overlap Corrections," *Hewlett-Packard HP-67/97 User's Library Solutions in Physics*.

APPENDIX I

LEAST SQUARES FIT TO A STRAIGHT LINE

The method of linear least squares is used extensively throughout molecular biology and biochemistry. A great deal of effort on the part of theoretical chemists and biochemists has been expended to develop linearizations of nonlinear phenomena. Linearized data are highly desirable since statistical theory is on much firmer ground with respect to linear than to nonlinear processes. In some instances, the data for a particular process under investigation (ligand binding, for example) can be linearized over only a portion of their complete range. In these cases more thorough examination may require nonlinear least squares analysis. We provide here the basic concepts of linear least squares analysis (linear regression and the equations which have been derived to determine the most probable values of the slope and intercept).

Suppose that we have data consisting of pairs of measurements (x_i, y_i) of an independent variable x and a dependent variable y. We. wish to fit the data to an equation of the form

$$y = a + bx \tag{1}$$

by determining the values of the coefficients a and b so that the differences between the values of our measurements y and the corresponding calculated values $y = f(x_i)$ given by equation 1 are minimized. We cannot determine the coefficients exactly with only a finite number of observations, but we do want to extract from these data the most probable values for the coefficients.

For the purposes of this discussion we define χ^2 as the sum
of the squares of the deviations

$$\chi^2 = \sum \left[\frac{1}{\sigma_i^2} [y_i - f(x_i)]^2 \right] \tag{2}$$

where σ_i is the standard deviation of the ith value of y, and
$f(x_i) = a + bx_i$. Our method for finding the optimum fit to the
data is to minimize this weighted sum of squares of deviations
χ^2, that is to find the *least squares fit*.

To find the values of the coefficients a and b which yield
the minimum value for χ^2, the method of calculus is used to minimize the function with respect to more than one coefficient. The
minimum value of the function of χ^2 of equation 2 is one which
yields a value of zero for both partial derivatives with respect
to each of the coefficients

$$\frac{\partial}{\partial a} \chi^2 = \frac{\partial}{\partial a} \left(\frac{1}{\sigma^2} \sum (y_i - a - bx_i)^2 \right)$$

$$= \frac{-2}{\sigma^2} \sum (y_i - a - bx_i) = 0$$

$$\frac{\partial}{\partial b} \chi^2 = \frac{\partial}{\partial b} \left(\frac{1}{\sigma^2} \sum (y_i - a - bx_i)^2 \right) \tag{3}$$

$$= \frac{-2}{\sigma^2} \sum [x_i(y_i - a - bx_i)] = 0$$

where we have for the present considered all of the standard
deviations to be equal, that is, $\sigma_i = \sigma$. These equations are rearranged to yield the so-called *normal equations,* a pair of simultaneous equations

$$\sum y_i = \sum a + \sum bx_i = aN + b\sum x_i$$

$$\sum x_i y_i = \sum ax_i + \sum bx_i^2 = a\sum x_i + b\sum x_i^2 \tag{4}$$

where we have substituted N for $\Sigma(1)$ since the sum runs from
$i = 1$ to N. Equations 4 can be solved for the coefficients a and
b for which χ^2, the sum of the squares of the deviations of the
data points, is a minimum. Solving for a and b results in the
following equations

$$b = \frac{\sum x_i y_i - (\sum x_i \sum y_i / N)}{\sum x_i - [(\sum x_i)^2 / N]}$$

$$a = \frac{\sum y_i}{N} - b \frac{\sum x_i}{N}$$

(5)

If the uncertainties due to instrumental errors and the like are not equal throughout, it is necessary to reintroduce the standard deviation from equation 2 as a weighting factor into equations 3 and 4. Instead of minimizing the simple sum of the squares of deviations as in equations 3, we weight each term of the sum in χ^2 according to how large or small the deviation is expected to be at that point before summing. An example of this method of weighted least squares is provided by the program in Section 4D.

We must now ask the question whether, indeed, there exists a relationship between the variables x and y. "Correlation" is the term used to describe this relationship. The linear correlation coefficient r derived from correlation probability theory will indicate quantitatively whether or not we are justified in assuming even the simplest linear correspondence between the two quantities. The equation for the linear correlation coefficient is

$$r = \frac{N\sum x_i y_i - \sum x_i \sum y_i}{\left[N\sum x_i - (\sum x_i)^2\right]^{\frac{1}{2}} \left[N\sum y_i^2 - (\sum y_i)^2\right]^{\frac{1}{2}}}$$

(6)

The value of r ranges from 0, when there is no correlation, to $\pm$ ±1, when there is complete correlation. The TI calculator programs in this book give the correlation coefficient, whereas, in most instances, the HP calculator programs give the value r^2, which is usually referred to as the coefficient of determination. This number will always be a positive number between 0 and 1.

Although the statistic most often quoted for data analyzed by least squares procedures is the correlation coefficient, there is some doubt as to the value of the correlation coefficient as a valid indicator of the quality of the least squares fit. Other statistics such as the standard error of the estimate and the standard deviations of slopes and intercepts are quoted less frequently. The equations for these statistics are provided here in a convenient form to use in conjunction with most of the least

squares curve fitting programs in this book.
The standard error of the estimate (of y on x):

$$S_{y \cdot x} = \left[\frac{\sum y_i^2 - a\sum y_i - b\sum x_i y_i}{N - 2} \right]^{\frac{1}{2}}$$

The standard deviation of the intercept:

$$S_a = S_{y \cdot x} \left[\frac{\sum x_i^2}{N\left(\sum x_i - (\sum x_i)^2/N\right)} \right]^{\frac{1}{2}}$$

The standard deviation of the slope :

$$S_b = \frac{S_{y \cdot x}}{\left[\sum x_i^2 - (\sum x_i)^2/N \right]^{\frac{1}{2}}}$$

Note that N must be a positive integer > 2.

FURTHER READING

1. Philip R. Bevington (1969), *Data Reduction and Error Analysis for the Physical Sciences*, McGraw-Hill Book Company, New York, Chapter 6 and 7.

2. R. B. Davis, J.E. Thompson, and H.L. Pardue (1978), "Characteristics of Statistical Parameters Used to Interpret Least-Squares Results", *Clinical Chemistry 24*, 611-620.

PHYSICAL CONSTANTS AND CONVERSION FACTORS

PHYSICAL CONSTANTS

Avogadro's number
$$N = 6.0238 \times 10^{23} \text{ gm/mole}$$

Boltzmann's constant
$$k = 1.3804 \times 10^{-23} \text{ joule/}^\circ\text{K}$$
$$= 1.3803 \times 10^{-26} \text{ erg/}^\circ\text{K}$$
$$= 8.617 \times 10^{-5} \text{ eV/}^\circ\text{K}$$

Charge on electron
$$\varepsilon = 4.802 \times 10^{-10} \text{ esu (statcoulomb)}$$

Curie
$$\text{Ci} = 3.7 \times 10^{10} \text{ dps}$$
$$= 2.22 \times 10^{12} \text{ dpm}$$

Dielectric constant of water
$$C = 78.54 \ (25^\circ\text{C})$$

Density of water
$$\rho_w = 0.9982 \text{ gm/cm}^3 \ (20^\circ\text{C})$$

Faraday ($N\varepsilon$)
$$F = 96{,}487 \text{ coulombs/mole}$$
$$= 96{,}487 \text{ joules/mole}\cdot\text{V}$$
$$= 23.06 \text{ kcal/mole}\cdot\text{V}$$

Gas constant ($N\kappa$)
$$R = 8.317 \text{ joules/gm-mole}\cdot{}^\circ\text{K}$$
$$= 8.316 \text{ ergs/gm-mole}\cdot{}^\circ\text{K}$$
$$= 1.987 \text{ cal/gm-mole}\cdot{}^\circ\text{K}$$

Gravitational acceleration
$$G = 980.616 \text{ cm/sec}^2 \text{ (sea level, } 45^\circ \text{ lat.)}$$

349

Partial molal volume of $\bar{v}_w$ = 17.984 cm /mole (25°C)
water

Planck's constant h = 6.625 × 10^{-37} joule·sec
                                = 6.625 × 10^{-27} erg·sec
                                = 4.134 × 10^{-15} eV·sec

Surface tension of water σ_w = 71.97 dynes/cm

Velocity of light in a C = 2.9979 × 10^{10} cm/sec
vacuum

                          RT = 592 cal/mole (25°C)
                               = 583 cal/mole (20°C)
                               = 24.47 liter-atm^{-1} (25°C)
                               = 22.41 liter-atm^{-1} (20°C)

                      RT/F = 29.6 mV (25°C)
                               = 25.3 mV (20°C)

CONVERSION FACTORS

Angstrom   = 0.1 nm
                                = 10^{-8} cm

Atmosphere   = 760 mm Hg (sea level)
                                = 0.1013 joule/cm^3
                                = 1.013 × 10^6 dynes/cm
                                = 1.013 bars

Calorie   = 4.184 joules

Coulomb   = joules/V

Dyne   = gm-cm/sec^2

Electron volt   = 1.602 × 10^{-12} erg

Erg   = dyne-cm
                                = 2.390 × 10^{-11} kcal
                                = 6.242 × 10^{11} eV

Farad   = coulombs/V

Joule   = 0.239 cal
                                = coulomb/V
                                = 10^7 ergs

Liter-atmosphere $= 24.2$ cal

Statcoulomb $= 3.336 \times 10^{-10}$ coulomb

SELECTED THERMODYNAMIC DISSOCIATION CONSTANTS*

Buffer	pK_a (25°C)	dpK_a/dt
Sulfuric acid (pK_2)	1.96	0.015
Phosphoric acid (pK_1)	2.15	0.0044
Glycine (pK_1)	2.35	-0.002
Citric acid (pK_1)	3.13	-0.0024
Glycylglycine (pK_1)	3.14	0.0
Malic acid (pK_1)	3.40	--
Succinic acid (pK_1)	4.21	-0.0018
MES [2-(N-morpholino) ethanesulfonic acid] (20°C)	6.15	-0.011
Acetic acid	4.76	0.0002
Citric acid (pK_2)	4.76	-0.0016

* D. Perrin and B. Dempsey (1974), *Buffers for pH and Metal Ion Control*, John Wiley & Sons, Inc., New York, pp.157-163.

Buffer	pK_a (25°C)	dpK_a/dt
Malic acid (pK_2)	5.13	--
Succinic acid (pK_2)	5.64	0.0
Cacodylic acid	6.27	--
Carbonic acid (apparent pK_1)	6.35	--
Citric acid (pK_3)	6.40	0.00
Bis-tris	6.46	--
ADA [N-(2-acetamido) iminodiacetic acid] (20°C)	6.62	-0.011
PIPES [piperazine-N,N'-bis (2-ethanesulfonic acid)] (20°C)	6.80	-0.0085
ACES [N-(2-acetamido)-2- aminoethanesulfonic acid] (20°C)	6.88	-0.020
Imidazole	6.95	-0.020
BES (20°C)	7.17	-0.016
MOPS (20°C)	7.20	--
Phosphoric acid (pK_2)	7.20	-0.0028
TES (20°C)	7.50	-0.020
HEPES (20°C)	7.55	-0.014
Tris [tris(hydroxymethyl) aminomethane]	8.06	-0.028
Tricine (20°C)	8.15	-0.021
Glycylglycine (pK_2)	8.25	-0.025

Buffer	pK_a (25°C)	dpK_a/dt
Bicine (20°C)	8.35	−0.018
Histidine	9.18	−−
Boric acid	9.23	−0.008
Ammonia	9.25	−0.031
Ethanolamine	9.50	−0.029
CHES (20°C)	9.55	−−
Piperazine (pK_1)	9.81	−0.022
Phenol	10.00	−0.009
Carbonic acid (pK_2)	10.33	−−
CAPS (20°C)	10.40	−−
Piperidine	11.12	−0.031
Phosphoric acid (pK_3)	12.33	−0.026
Guanidine	13.60	−0.026

APPENDIX IV
OTHER SOURCES OF CALCULATOR INFORMATION

HEWLETT-PACKARD HP-67/97

1. *PPC Journal,* formerly *65 Notes*
   Richard J. Nelson, Publisher
   2541 W. Camden Pl.
   Santa Ana, CA   92704

2. *HP-67/97 User's Library*
   Hewlett-Packard
   1000 N.E. Circle Blvd.
   Corvallis, OR   97330

3. *Hewlett-Packard HP-67/97 User's Library Solutions Books.*
   Collections of programs submitted to User's Library and published in convenient booklet form, grouped according to
   specialty.

4. *Hewlett-Packard HP-67/97 Application Pacs,* for example:
   Clinical Lab and Nuclear Medicine Pac
   Stat Pac I
   Math Pac I

TEXAS INSTRUMENTS TI-58/59

1. *PPX-59--software exchange* sponsored by Texas Instruments
   Texas Instruments
   P.O. Box 53
   Lubbock , TX   79408

2. *Texas Instruments Plug-in Solid State Software Library*
    (includes CROM and instruction manual), for example:
    Master Library Module (included with TI-58/59)
    Applied Statistics
    Math Utilities Library
    RPN Simulator
    Leisure Library

3. *Texas Instruments Specialty Packettes*--collections of pro-
    grams submitted to *PPX-59* in convenient booklet form analog-
    ous to HP's User's Library Solutions Books

GENERAL

The following scholarly journals often contain articles and
programs for specific programmable calculators:

1. *Computer Programs in Biomedicine*
2. *Computers in Biology and Medicine*

The following two books offer useful suggestions for im-
provements in calculator algorithms and methods of analysis:

1. *Scientific Analysis on the Pocket Calculator* by Jon M. Smith,
    John Wiley & Sons, Inc., New York, 1975.
2. *Algorithms for RPN Calculators* by John A. Ball, John Wiley &
    Sons, Inc., New York, 1978.

APPENDIX V

HP-41C
COMPATIBILITY

Hewlett-Packard's recently announced HP-41C calculator combined with the optional card reader was designed so that most
prerecorded HP-67/97 programs can be read and executed on the
HP-41C. The card reader translates the HP-67/97 program instructions into special compatibility instructions that execute on the
HP-41C. You can also key in and use the compatibility functions
while the card reader is attached to the HP-41C. A complete
listing of all the card reader functions is given in the owner's
handbook. General instructions are also provided in the card
reader owner's handbook for properly reading HP-67/97 program
cards and executing the programs.

Most of the programs presented in this book will fit into
the basic memory of the HP-41C calculator with card reader
attached. Table 1 shows the memory requirements of all of these
programs. The basic HP-41C has 17 data registers and 46 program
registers at turn on.

The number of registers used and the number left after
reading the programs into calculator memory are indicated in
Table 1. Four of the programs require and additional memory
module before translation and execution can take place. If the
user of an HP-41C does not have access to an HP-67 or HP-97 and
does not have an additional memory module, it may still be
possible to use these programs by carefully inspecting them and
eliminating all nonessential program steps and/or registers.

TABLE 1

Section	Program	Registers Used*	Registers Left
1A	Buffers-I	12	25
	Buffers-II	17	20
B	Weak Acid/Base Titration†	51	50
C	Acid-Base Equilibrium	22	15
2A	Partial Specific Volume	12	25
B	M.W. from Osmotic Pressure	26	11
C	Intrinsic Viscosity	30	7
D	M.W. from Gel Filtration	24	13
E	Diffusion Coefficient-Width Method	35	2
3A	Routine Centrifugation Calculations	29	8
B	M.W. by Sedimentation Velocity	34	3
C	M.W. by Equilibrium Sedimentation	34	3
4A	Scatchard Plot	31	6
B	Hill Plot	28	9
C	First Order Kinetics	15	22
	Second Order Kinetics†	46	55
D	Michaelis-Menten Enzyme Kinetics†	45	56
E	K_m and V_{max} Single Progress Curve	34	3
F	Arrhenius Plot*	39	0
5A	Nernst Plot	27	10
B	Activity Coefficients	37	0
6A	Beer-Lambert Law	36	1
B	Multicomponent Spectroscopy	8	29
C	Radiative Fluorescence Lifetime	19	18
D	Fluorescence Polarization	9	28
E	Circular Dichroism	16	21
F	3 × 3 Matrix Operations†	46	55
7A	Radioisotope Decay	20	17
B	Isotope Overlap Corrections	18	19

*Assumes *SIZE* 026 except for Section 4F. Arrhenius Plot requires *SIZE* 024.
†Requires one additional memory module.

358

There are a few basic techniques the HP-41C owner can apply to make these programs fit without having to reprogram or even understand the details of the program code. A few of these techniques are described below.

1. *Balance memory.* The normal procedure for reading HP-67/97 programs is to have the HP-41C *SIZE* set to 026. This is essential if:

> a.   The 67/97 program uses all 26 data registers or,
> b.   The 67/97 program uses the I register (R25 in the HP-41C).

The trick here is to examine the 67/97 program and determine how many registers are being uses. Suppose, for example, you observe that the program uses R1, R2, R5, RA (R20), and RE (R24) for a total of six registers. Reassign the registers starting at R00. Make a table like the one shown in Table 2. This makes it easy to keep track.

TABLE 2

HP-67/97 Original Register		HP-41C Reassigned Register
R1	→	R00
R2	→	R01
R5	→	R02
R9	→	R03
RA (R20)	→	R04
RE (R24)	→	R05

Follow the program listing and carefully mark each *STO* and *RCL* number with the new assignment. Carefully key the program into the HP-41C using the new register assignments in place of the old being sure that the *SIZE* has been adjusted appropriately (006 in this example).

2. *Remove unnecessary steps.* The card reader adds steps during the translation process. For every letter label in the HP-67 program a numeric label is added before the letter label to speed

up execution. Since speed is less critical at this point, delete
all translated numeric labels ahead of all letter labels. The
HP-41C user may key the program in directly just using the letter
labels. Also replace all *PRTX* instructions with *STOP*. All programs which use the HP-67/97 summation registers in linear
regression routines will require that the HP-41C instruction
ΣREG 14 be executed during initialization. All *DSPn* instructions, where *n* is 0 to 9, could be deleted if necessary. Simply
set the display to a convenient mode and number of digits and
eliminate the "frills". One word of caution, however; in the
programs which use the *RND* function the preceding number of display digits set by the program is important. Don't remove the
DSPn in this case.

     Removal of the *P⇄S* is potentially the most dangerous. If
you understand the operation of the program this instruction can
be eliminated by simply reassigning secondary registers as
described in step 1. It is not always clear which register is
being addressed if there are multiple *P⇄S* instructions and con-
voluted logic so be careful. (See Weak Acid/Base Titration
program in Section 1B.) Also be careful with the *DSZ, DSZI, ISZ,*
and *ISZI* instructions. These cannot be directly replaced with
the HP-41C instructions *DSE* and *ISG*. The compatibility functions, *7P⇄S, 7DSZ,* etc., will have to be used instead.

3. *Split into multiple card programs.* Break the program into
two parts where appropriate. It should be possible to run the
programs properly without any further modification.

     If these techniques do not work, the HP-67/97 program may
have to be completely rewritten to utilize the maximum efficiency
of the HP-41C. The best alternative is to simply purchase an
additional memory module. These techniques were excerpted from
various articles written and published by Richard J. Nelson in
PPC Journal. See Appendix IV for the address of PPC.

INDEX

Date Due